A River Through Illinois

A River Through Illinois

PHOTOGRAPHS BY Daniel Overturf   WORDS BY Gary Marx

WITH A FOREWORD BY BILL KURTIS

SOUTHERN ILLINOIS UNIVERSITY PRESS   CARBONDALE

Depth gauge on disused barge, photographed near Creve Coeur. River Mile Marker 159.5, Christmas Day 2001.

CONTENTS

Maintenance bridge technician Gene Smania of the Illinois Department of Transportation, photographed on the northwest side of the Illinois Route 178/Utica Bridge. River Mile Marker 229.5, March 2004.

ACKNOWLEDGMENTS

A number of people have contributed to this project, and the authors would like to express their appreciation. Many kind individuals have shown their understanding and at times guided the authors on their journey. In compiling a list of those to thank, the risk of omission is always great; apologies and appreciation are extended to any victim of oversight.

Of special note are the following people who facilitated some of the circumstances that led to many of the photographs, interviews, and experiences contained in A River Through Illinois: Dora Dawson (Meredosia); Ron and Vikki Jones (Chillicothe); Kenny and Betty DeFord (Bath); Gene Smania (La Salle) of the Illinois Department of Transportation; Luke Moore and Tom Flowers with Western Kentucky Navigation; Doug Blodgett, Joanne Skoglund, and Tharran Hobson at the Nature Conservancy Great Rivers Area; Bill Recktenwald (Karbers Ridge); and Jerry Arp (Chicago).

The following organizations also assisted in granting access to places and personnel: Metropolitan Water Reclamation District of Greater Chicago; Chicago Police Department; U.S. Army Corps of Engineers; Friends of the Illinois River; Illinois Corn Growers Association; Illinois Department of Transportation; Illinois Department of Natural Resources; the Nature Conservancy; and Western Kentucky Navigation. Through a Fellowship Award, the Illinois Arts Council assisted in the making of the photographs.

Lyrics from "Long Hot Summer Days" are reprinted by permission of John Hartford Estate. Words and music by John Hartford. Lyrics from "Help Me Make It through the Night" are reprinted by permission of Hall Leonard Corporation. Words and music by Kris Kristofferson, copyright 1970, renewed 1998 Temi Combine Inc. All rights controlled by Combine Music Corp. and administered by EMI Blackwood Music Inc. All rights reserved. International copyright secured.

The authors thank Julie Bush for her skillful editing and Pamela Kelley for her patience and advice. The creation of the images, over the course of many years and for many purposes, has been a collective endeavor. Of particular note are Steve Nedza (Chicago), who elevated the images in his Lambda versions of the negatives; Laura Lange, Scott Price, and Brian Matsumoto, who helped with equipment support; Southern Illinois University at Carbondale graduate students Josh Sanseri, Carla Cioffi, Polly Chandler, and, especially, Dimitris Skliris, who made work prints of the images over the years; Lyle Fuchs, who provided laboratory and hardware support; David and Jane Gilmore, who gave unending assistance in the framing of the photographs, and especially David for being a teacher, mentor, and good friend who has provided knowing advice for nearly thirty years; Heather Lose for her design skills and patience; Ann Dodge for the digital preparation of the book versions; and Kelvin Sampson and Julie Barr of Dickson Mounds State Museum for their excellent work on the presentation and gallery arrangements for the April 2004 exhibition.

Finally, both of the authors extend their heartfelt appreciation to their parents—Alfrieda and the late Leonard Marx and Wayne and the late June Overturf—for their lessons, for their guidance, and for the examples they set with their lives.

Abandoned rope on a
wooden mooring cell near
Peoria. River Mile Marker
163, Christmas Day 2002.

FOREWORD   *Bill Kurtis*

There was a time when rivers defined America.

They were the first portals into the new land, an amusement ride past natural wonders taking explorers and traders into a vast new world.

They were the lifeblood of a new country. No city of any note lasted long without a river to quench the thirst of its people, take away their waste, and provide power for their mills.

Rivers were the silent partners of America, as necessary as the air we breathe, as eternal as time itself.

Eight thousand years before Europeans discovered America, indigenous tribes used the rivers to build a system of commerce that connected the oceans and led to a city of giant earthen mounds as large as Egypt's pyramids at a place called Cahokia, near where the great waterways met. Today, the mounds are still there, overlooking St. Louis from the Illinois banks of the Mississippi.

One hundred years after the Declaration of Independence, the river system continued to be the primary means of transportation, and it seemed like the whole country was traveling on it. The dazzling promise of free land was in their heads as immigrants floated the Ohio River, then changed to the Mississippi and the Missouri as casually as their descendants would change trains a few years later.

In this migration across the continent, more and more homesteaders settled along the smaller rivers, like the Illinois, where they created cultures of their own. Generations would rear families and make a living, never having to leave the river culture. They would hug the riverbanks like amphibians that were strangely connected to the current, as if their hearts beat in rhythm to the flow.

But over the years, as superhighways leapt over them in a single bound and commerce disappeared in vapor trails high above, rivers lost their dominance. Gone were the ferries and gristmills, the roadhouses and trotlines. History grew over the river culture like weeds over abandoned fishing shacks.

Fortunately, all is not lost. We have a record of what once was and what remains of this important part of our history, as seen through the eyes and images of Daniel Overturf and Gary Marx. Their photographs and words bring the culture alive so that we can feel it, see it, and better understand it. This is an important book.

Those who fear that the world of the river will be lost before the story is fully told need fear no longer. *A River Through Illinois* cements its place in history. It is part of America's narrative, of people who live their lives on the river and help give Illinois its soul.

A slow-moving Amtrak train crosses
the Chicago River near Canal Street,
between Cermak Road and 18th Street.
South Branch of the Chicago River Mile
Marker 323.5, March 2000.

*Barge wheelhouse pilot
entering Marseilles Lock;
two-minute exposure.
River Mile Marker 244.5,
August 2001.*

Introduction

The story of this river, the Illinois Waterway, cannot be told by a single voice. It takes a chorus.

Along its length—330 miles from Chicago to Grafton—individual voices can be heard telling distinct and full stories, each but a piece of the whole. This is a river of stories, told in a mosaic of sights and sounds.

Some of the voices are loud. There is a riotous clang, a cacophony of shouts—the first mate barks, bells ring, and metal bangs. Under a bridge, a machinist curses a bilge pump, and it finally kicks in.

There are quiet voices, too. Lovers whisper in boats. And the elders muse, recalling a time when there were endless tomorrows giving promise to their todays.

The story of the river is told, as well, by a voice that belongs to no one. It can be heard in the invisible slip-slap of water, in a wisp of wind, and in the startled flutter of wings. It can be heard in the duck blind, in that swollen moment just before dawn when even the dog holds its breath. It is a voice that is sometimes seen and felt.

We have a bond with our rivers. We refer to them with familiarity, dropping the formality imposed by the word "River." They are simply the Ohio, the Wabash, the Missouri. We give them nicknames: Big Muddy, Old Man, the Miss. This vernacular speaks to the affinity we have for these waters that both bless us with their bounty and forsake us in those heartless days of flood and rage.

We have tried to tame the rivers, channeling them and binding them within levees and dams, but they only rise up and remind us that there is a greater power.

The Illinois Waterway comprises all of the Illinois River; parts of the Des Plaines, Chicago, Calumet, and Little Calumet rivers; and two canals—the Chicago Sanitary and Ship Canal and the Calumet–Sag Channel. The whims of geography and political boundaries have placed the waterway completely within the borders of the state that gives it its name. More than 90 percent of the state's population lives in its watershed. This waterway very much belongs to Illinois.

But it also belongs to the world, a vital link in the food chain of commerce. The state's abundant agricultural products and manufactured goods are shipped on these waters to markets across the globe. Its major tributaries and manmade canals reach Lake Michigan, forming a continuous route between the

Great Lakes and the Mississippi River, between the North Atlantic and the Gulf of Mexico. You can reach anywhere in the world by sailing in either direction on the Illinois River.

These waters will take you to the city, through the shadows of skyscrapers in the belly of Chicago. They will carry you to the industrial region above Joliet and to the manufacturing heartland that is Peoria. They will take you below the bluffs, through the prairies and forests that hint of Native Americans, Marquette and Jolliet, dugouts and canoes.

This is more than a journey in geography. All the way to Hardin and Grafton, the Illinois will take you back to the time of flatboats and steamboats to the days before locks and dams and levees. This is a journey in time, too.

This project, a partnership of words and images, is an attempt to tell the story of the waterway—the Illinois River and the arms that reach Lake Michigan. It is in many respects incomplete, for there are so many facets to the tale. This is, rather, only one of its stories. It is a composite, a mosaic, a time-exposure, and it is our intention that those who look through the lens of this book will see this river that flows through Illinois in new light and that they will gain a fresh appreciation for it and the people whose combined voices tell this story.

A river level indicator from an earlier time and cable-wrapped mooring cells, looking downstream to the Cedar Street Bridge and the ADM plant, Peoria. River Mile Marker 162, November 2000.

Part One

For every day I work on the Illinois River,

Get a half a day off with pay,

On a towboat making up barges

On a long hot summer day.

—JOHN HARTFORD, "Long Hot Summer Days"

Looking west into the wilder side of the Chicago Portage National Historic Site Forest Preserve District, located on Harlem Avenue near Interstate 55. Near Chicago Sanitary and Ship Canal Mile Marker 314, July 2003.

CHAPTER ONE

Portages

His boots are caked with mud, his leggings torn and his shirt damp with sweat. Under his broad hat, a bead of perspiration forms on his brow as his eyes peer beyond the marsh. His nostrils flare. This place reeks of onion.

Louis Jolliet has spent months in the wild, and now on this low hill just north of the marsh, he knows he and the Jesuit have arrived at a very special place. It is late summer, 1673.

Jolliet, a twenty-eight-year-old fur trader and cartographer, had led Father Jacques Marquette, age thirty-six, and a handful of French voyageurs from what is now Canada through Wisconsin and down the Mississippi River. They had heard that the river might lead to the Pacific Ocean. They discovered that it did not.

On their return, the small expedition—two birch canoes, a total of seven men—had befriended a party of Illinois Indians who showed them a shorter route to the great lake and back to Canada.

They canoed up the slow, narrowing Illinois River, named for the people who were native to the land. They entered the Des Plaines, one of its two main tributaries. And then at a point not far from where the river hooked to the north, the party paddled up a little

creek on the eastern bank. When they could go no further, they pulled and pushed their canoes through the grasses and sand that skirted the shallow marsh.

Jolliet wipes his brow and peers ahead. Not far away, another stream leads east to the great lake and back to Canada. The significance of this place is immediately clear.

The lake was so close to the tributaries of the Mississippi River that a short canal—a few miles in length, perhaps—could link those two waters. Such a canal, Jolliet realized, would eliminate the need to portage and would open up the entire midsection of North America to exploration and trade.

Jolliet was correct about that, but it would be 150 years before the first shovel was lifted on the canal, and by then the world would be a vastly different place.

Today, a hulking steel sculpture marks the spot where Jolliet and Marquette portaged from the Des Plaines to the Chicago River. Its industrial block-like figures depict the two explorers and an Illinois Indian pushing a canoe. It stands in a nondescript park off Harlem Avenue at the Stevenson

The cross commemorating Jolliet and Marquette's entering the Illinois River, on Illinois Route 100, west of Grafton, near Pere Marquette State Park. Near River Mile Marker 7, November 2002.

Expressway, Interstate 55, just outside Chicago's city limits.

The park does not attract many visitors. It is a quiet bed of grass in the midst of the urban hustle. But it is not so quiet that the bump and grind of delivery trucks and automobiles on Harlem Avenue or the faster rhythms of the expressway can be ignored. Across the street to the east lies the local office of an oil company, and just beyond that is a small petroleum tank farm.

Trails lead west from the statue, snaking off toward a body of water, a pungent algae-breeding pool that once upon a time connected to the Des Plaines River. Signs are posted to trees here and there, informing the hiker that this is the site of the Chicago Portage.

The landscape in no way resembles the land Jolliet and Marquette saw. The rivers have been channelized, relocated, filled in. The marsh, which came to be known as Mud Lake, was drained long ago. Only the notes of historical societies and a visitor's mind can conjure an image from September 1673, when Jolliet stood here, right here, on this very spot, and strained his eyes beyond the marsh and dreamed of a canal.

GLACIAL BEGINNINGS

The spot Marquette and Jolliet crossed is a geological oddity. It is a low-rising continental divide of sorts. The waters on one side flow east into Lake Michigan, eventually finding the Atlantic Ocean, while the waters on the other side flow west, ending up in the Mississippi River and the Gulf of Mexico. The proximity of rivers and the low barrier hills made this the ideal place to portage and eventually canal.

That mid-continental divide, formed by the slow dance of glaciers thousands of years ago, is barely perceptible today, lost in a landscape of asphalt and concrete. But it is there. The words "Heights" and "Hills" occur often in the names of suburbs and villages around the city of Chicago. Towns named Highland Park, Park Ridge, Blue Island, and Summit bear testimony to the undulating ground one does not even notice today.

The glaciers had carved out the basins for the Great Lakes, and they left a series of moraines along the southern and western shores of Lake Michigan. Those low-rising ridges run from Wisconsin to Indiana, and they retained the meltwater from the receding glaciers. Although it was slow and relentless, the glacial retreat was not a clean, smooth affair. At times, the glacial melt spawned sudden, catastrophic events.

When the last glacier receded to the northeast, about 14,000 years ago, the meltwaters would occasionally burst through the retaining ridges, sending violent torrents to the lower land to the west and south and carving what is today the Illinois River valley.

The Illinois River is curiously shaped. From the confluence of the Kankakee and Des Plaines rivers, it flows almost directly west, and then the river abruptly hooks left and flows south for about 200 miles.

At one time, geologists believe, the Mississippi River occupied the lower Illinois River valley, but during the last glacial age a large mass of ice dammed it off and rerouted it to the west. Once the glacier receded, the newly formed Illinois River followed the path of least resistance, which was the wide valley already carved by the old Mississippi River. As the glaciers dissipated, the lake levels dropped and the rivers settled into courses that they roughly follow today.

THE LAY OF THE LAND

One of the major rivers of Illinois, the Des Plaines, rises in southern Wisconsin and flows almost due south, running parallel to the Lake Michigan shore for nearly fifty miles. For most of that run, it is barely six miles away from the lake. A glacial moraine separates the river from the lake, and southwest of the present city of Chicago, the Des Plaines River turns west and heads toward the Mississippi River.

Also running south, closer to the lakeshore, is the North Branch of the Chicago River. For about twenty miles, the North Branch and the Des Plaines run side by side, separated by only the moraine. At one point, the two rivers are less than a mile apart. The North Branch and the opposite-flowing South Branch of the Chicago River meet head-on at Wolf Point—where today Lake Street and Wacker Drive intersect in the city—and the merged river turns east. In less than a mile it flows into the lake.

The Chicago was a slow-flowing stream. Both branches drained the low-lying swamps and marshes between the lake and the moraine. The South Branch, which received the explorers' canoes in 1673, sprang from the marsh called Mud Lake. That lake had the distinction during times of high water of draining in two directions because it straddled the mid-continental divide. As the Illinois Indians already knew—and later European explorers would discover for themselves—it was possible when the water was in flood stage to paddle between the Des Plaines and the Chicago without having to portage. Using the records of Marquette, Jolliet, and subsequent explorers, historians have speculated that the water must have been relatively high when those first two Europeans came through, because those who followed found the portage much longer and exceedingly more difficult.

All of these things—the rise and fall of the rivers, the seasonal changes that occur to the marshes and Mud Lake—were known to the Illinois Indians. And if the party of Native Americans that led Jolliet to the Chicago portage site knew what was in his mind as he scanned the landscape that day in September 1673, it is unlikely they would have shared his enthusiasm. For the Illinois Indians and their way of life, vast changes were coming—and coming quickly.

The Illinois were among the last of the Native American tribes to inhabit the Illinois River valley, but various people had lived in this region for thousands of years. Anthropologists and archaeologists have identified three distinct ages of human habitation here, tracing those ages almost as far back as the glacial period.

The valley was an attractive place in which to live. It was rich in wildlife, and the soil was fertile. The glacial melt had formed a river that was slow and benign. A broad floodplain allowed the river to breathe in times of high water. An incredibly diverse system of lakes and wetlands lined the valley and provided habitat for fish, fowl, mussels, and mammals.

This was the river as Marquette and Jolliet found it.

There is an often-quoted passage from Marquette's journal in which he describes the Illinois River just after they entered it from the Mississippi. "We have seen nothing like this river that we enter," he wrote, "as regards its fertility of soil, its prairies and woods, its cattle, elk, deer, wildcats, bustards, swans, ducks, parroquets, and even beaver. There are many small lakes and rivers. That on which we sailed is wide, deep and still."

True, they had seen nothing like it, but by the time they left the water—nearly 300 miles upstream—they were proposing a plan that would change it forever.

MANIFESTED DESTINY

As it turns out, Louis Jolliet was a bit shortsighted. He had estimated that a canal of only "half a league," roughly a mile and a half, would effectively link the Chicago and Des Plaines rivers. But the canal eventually built would bypass the Des Plaines River entirely, linking directly to the Illinois River almost 100 miles away.

The Illinois & Michigan Canal would take twelve years to complete, officially opening in 1848. It was,

for the most part, sixty feet wide and maintained a depth of six feet. Beyond its connecting lock in Chicago, the 141-foot descent to the Illinois River at La Salle was controlled with fifteen locks.

Jolliet may have miscalculated the magnitude of the project, but he was correct in at least one thing: the canal would have a profound effect on the territory. During the course of the canal's construction, the entire Illinois River valley was inexorably altered.

Even before a shovel had been turned, the canal's potential was palpable. The success of the Erie Canal decades earlier had spawned a canal-building craze in the nation, and the I&M Canal project attracted thousands of laborers—many of whom were recent European immigrants—and hundreds of investors and land speculators.

As earth was moved, coal was discovered, as was limestone, silica, and a high-quality clay. These became the building blocks for the cities and towns along the canal route. Willow Springs, Lemont, Lockport, Joliet, Marseilles, Ottawa, Utica, and La Salle all trace their growth—if not their origins—to the canal. So does Chicago.

Because of the canal, Chicago was transformed from a frontier settlement to the nation's greatest inland port. The population of the city jumped from a few hundred to more than 20,000 in 1848, when the canal opened. Following the success of the canal, the railroad finally reached Chicago in the mid-1850s, and by 1860, more than 100,000 people lived in the city. By 1880, the population was half a million; by the time the Columbian Exposition opened in 1893, more than a million people lived in the city that the little canal had built.

But by then the canal was barely in use—mortally wounded in the war with the railroad.

Today, the canal is enjoying a second life as a tourist destination. Its history is preserved along the Illinois & Michigan Canal National Heritage Corridor, which links towns, historic sites, and trails along the route. For much of the way, a bicycle and hiking trail follows the old towpath.

Some of the I&M locks are well preserved in the corridor. An aqueduct, which carried the canal over the Fox River near Ottawa, has been restored. Fine historic buildings from the canal days can be seen at Channahon, Lockport, and other places. Museums and visitor centers along the way explain the canal's significance.

And running parallel to the canal is the legacy it spawned: the working canal today, the Illinois Waterway.

A Link to the Ancients

Dickson Mounds

If he had to pick one moment in his life, if he had to choose one image from the vast gallery in his memory, it would be this: Costa Rica. Thirty years ago. And he would be overlooking an old Spanish bridge.

Duane Esarey was just out of high school and was being shepherded around the Central American nation. Above that stone structure, someone pointed to it and said, matter-of-factly, "That bridge was built in 1550."

That fact, at that moment, turned on the light for him. That's when Duane Esarey knew something was missing.

"I grew up in central Illinois. I thought I knew the ground pretty well," he says. "I knew something must have happened in Illinois in the 1500s, but I didn't know a damn thing about it."

That moment in Costa Rica was an epiphany. And from that point overlooking the bridge built by the Spaniards in the sixteenth century, Esarey started on a path that would lead him back to Illinois, to the Illinois River valley, to Dickson Mounds State Museum.

Today Esarey is the assistant curator of anthropology at the museum, an educational and research facility near Lewistown, and the work he's done there has helped expand our understanding of the people who lived in the valley hundreds and thousands of years ago.

For Esarey, the study of civilizations is not just a look into the dark past; it is not a pursuit intended simply to satisfy some personal curiosity. It is both of those things, yes, but there is more to it. Anthropology for him is a means to provide perspective on the human condition today. Without a current context, he says, his study would be hollow.

That is why, while bent over a puzzle of pottery pieces thousands of years old, he is focused on what's happening today, just outside that door, down the hill and across the bottomland. He is focused on the river.

Assistant Curator of Anthropology Duane Esarey of Dickson Mounds State Museum with ancient pottery shards he found along the river, photographed on the east bank of the Illinois River, upstream of Havana. River Mile Marker 123, November 2001.

Dickson Mounds State Museum is located at an ancient Indian site on the western bluff overlooking the river, adjacent to the Nature Conservancy's Emiquon site. Across the river are Havana and the vast Chautauqua National Wildlife Refuge. There is a lot of activity in this particular section of the valley, but that's the way it has been for thousands of years, since the glacial age.

The fertile soil and an abundance of wildlife attracted humans. The remains of their villages dot these hills, and the earthen mounds they built still stand as evidence of the longevity of their civilization.

"Dickson Mounds is important," Esarey says, because "it does a good job of explaining how we fit in culturally. As you get more information, you start seeing how things are tied together." He has a specialty in prehistory and cultural adaptation, but the more he discovered, the more his study reached into other areas. All of the sciences are linked. He speaks of "the glee of understanding" and "this wonderful net of how things fit together."

One of the characteristics that make the Illinois River unique is the fact that most of it has been in the same channel for nearly 3,000 years. In times of flood, the backwater lakes and sloughs in the wide valley would fill, and tributaries would back up. When the valley gradually released the floodwater, a new chute might be cut here and there, but the river primarily would retreat to its old channel.

This fact was driven home in 1988, when Esarey and his colleagues made a remarkable discovery that led to the collection of "almost a full set of archaeological evidence for the past 3,000 years."

That year the entire Midwest was hit by a tremendous drought. Water levels dropped to fifty-year lows. Along the Mississippi and Missouri rivers, century-old steamboats, which had been buried in river mud, were baring their bones along the banks.

Using a fragment of previous research and following a hunch, Esarey and a team of volunteers embarked on a survey of the banks of the Illinois River. Working in several boats and walking the shoreline, the crews scoured the area for signs of previous civilizations.

They found one site, then another—and another. By the time they were done, they had documented more than 200 sites along a 100-mile stretch, from Naples to Peoria. Some of those sites were the remains of mile-long villages.

Today, Esarey's lab at Dickson Mounds is full of artifacts from that expedition. Drawers and boxes on tabletops are filled with pottery and tools, animal bones and chunks of charred wood, some of it bagged, all of it tagged. These pieces tell a story, he says, that "blew the roof off of what we knew" about life along the river over the past three millennia. And it also provided a perspective of the river as it exists today.

Those civilizations thrived because the river was benign. It was allowed to swell and then drain. There was a natural rhythm. It's not the same today.

"There are consequences to what we do," Esarey says. "We tend to not look at that. The basic nature of our species is . . . we are manipulators and pattern recognizers. We do things because we can. We drain wetlands and dump sewage to solve immediate problems and accomplish particular goals."

In the past 100 years, we have engineered all sorts of short-term solutions to our problems, he says, and now we are paying the long-term price.

The challenge—to reinvigorate the Illinois River, to help it do its work again—is daunting. The siltation problem alone is monumental. But there is also the chemical runoff, the flooding, the invasive species. None of these is far from the attention of the man who has devoted his life to looking backward: the anthropologist with an eye on the future.

"I suppose the next step for me is to move into a cave. I'm always on the verge of giving up," Esarey jokes, but then he grows serious. "I used to be really pessimistic, you know, in the '70s. I was worried that there's no way this was going to work out. But I had underestimated the resiliency and vigor of life."

His point is that we got ourselves into this mess; we can get ourselves out. We engineered the injuries to the river; we can engineer a solution.

"I don't have any answers. I just have pretty strong opinions of what we need to be. We need to be aware. Unless we know ourselves, we won't do anything right."

And if we don't?

"Well," he laughs, "we haven't messed up anything in Illinois that the next glacier isn't going to set straight."

Potowatomi Living History

Isle a la Cache Museum

Turtle Dove doesn't look like a Potowatomi Indian. She doesn't look French, either. But at the Isle a la Cache Museum on the Des Plaines River near Romeoville, visitors are encouraged to use their imaginations.

Ann Marie Van, an interpreter at the site operated by the Will County Forest Preserve, is Turtle Dove. She is dressed in traditional Indian skins, her hair pulled into braids. She has made an attempt to look the part, but her voice gives her away.

"I am half French," says Turtle Dove, in character and to the point. "My mother is Potowatomi. My father was a voyageur. I live in the Potowatomi village."

The museum's artifacts and exhibits try to show what life was like on these shores during the early eighteenth century, after the time of Marquette and Jolliet and while the cultures of the Native American and the French voyageur crossed. The museum does a credible job, as does Turtle Dove.

She explains the skins and hides and the tools as she walks visitors through the museum. Her tour places this portion of the river into a broad historical context. And as she talks, one wonders how a young woman named Ann Marie came to be Turtle Dove.

"We would trade these things with the voyageurs," she says. "We did not have trouble with the French as we did the English, because the French did not come to stay."

Isle a la Cache means "Island of the Hiding Place," the name coming from a legend that holds that the voyageurs stashed their furs and trading material on the island after hearing of dangers downstream. One version of that tale has the French traders hiding their goods on the island to avoid paying a toll to the tribe that controlled the river in this area.

The eighty-acre island does not much resemble an island today, especially when the water is low. The river channel has shifted and dropped. But when the water is high, the island becomes more prominent. The Des Plaines River runs parallel to the old I&M Canal and the Chicago Sanitary and Ship Canal at this point.

Turtle Dove explains the different types of French traders she and her people came into contact with in the 1700s. There were those who were officially sanctioned, having been granted one of the twenty-five annual licenses issued by the French government. They would travel with a group of hired voyageurs.

"The voyageurs were nothing more than truck drivers," Turtle Dove says. She doesn't explain how she knows what a truck driver is. "They did all the work, portaging the canoes and bringing the traders to the tribes. There were anywhere from eight to fifteen voyageurs with each trader."

And then there were the independent and somewhat illegal coureurs de bois, who traveled alone by canoe and fended for themselves. Both types of traders mixed easily with the Indian tribes.

"There was a lot of intermarriage between the Indian and the French," Turtle Dove says, a fact that is key to her story.

It takes a little prodding, but Turtle Dove gradually allows Ann Marie's own story to emerge. She came to this part of the country, to Plainfield, to live with her new husband, but tragedy struck their young marriage and she was widowed early. Fate had dealt her pain and cut her loose alone. She enrolled in school to see where that might lead, and it was in class that she discovered history.

"I started as a math major," the Oak Park native says. But after taking two required history courses, she discovered a passion for the region's early days. She eventually got her history degree at the University of St. Francis in Joliet and did her thesis on women in the fur trade. That brought her to the Isle a la Cache, where she served first as an intern, then as a volunteer, and now as a member of the staff. She suddenly catches herself, faintly uncomfortable about slipping out of character.

"The French would trade these metal pots to us," she says. Life is full of tradeoffs. "But we did not want them to cook in. We had used clay pots for a long time. We would use the metal to make arrow heads and tools. You see, most of the time, we wouldn't be looking at things in the same way as the French."

Cultures meet and meld and sometimes clash. Powerful forces alter lives. Those stories are told at the Isle a la Cache.

Metal sculpture at Isle a la Cache Museum, on the Des Plaines River. The museum is on Romeoville Road six miles north of where the Des Plaines takes over the Illinois Waterway from the Chicago Sanitary and Ship Canal, just downstream of the Lockport Lock and Dam. Parallel to Chicago Sanitary and Ship Canal Mile Marker 296, March 2004.

J. P. Hughes and Mabel Rounds fishing on south side of the I&M Canal in Morris. Behind them is the now-replaced Illinois Route 47 Bridge. River Mile Marker 263.7, July 2001.

Mabel and Hughes

Fishing Lesson at the I&M Canal in Morris

The bullheads will keep you busy. One pole or another will be jumping at any given time, and the bucket fills in a hurry. There will be fish in the pan tonight.

"We just come down here for the fun of it," Mabel says. She's quick to smile, especially when Hughes is around. It's plain to see they're sweet on each other. "Oh, we're just friends," she insists, waving away the inquiry as if it were a fly. But there's a blush trying to happen on her cheeks.

The two drove down to Morris from Chicago, a little more than an hour away, hoping to catch a few fish and enjoy the summer day. The fish weren't biting in the main channel of the Illinois, but over here in the still waters of the I&M Canal, it is a different story. Mabel Rounds and J. P. Hughes are not alone; a handful of other fishermen are catching bullheads, too. Occasionally a young boy, a local kid, wanders by and peers into the bucket, and Hughes gives him a tip or two.

"I don't know what it is," he says, "but whenever I go to the pond or the river or anywhere to fish, the kids they just come around. They must think, 'He's old, he must know how to fish.'" And Hughes lets out a laugh that comes from deep down. It resonates from somewhere back behind his heart. It becomes clear then that those kids come around for more than just a fishing lesson.

The Stones at Sag Bridge

Earning a Place to Rest at St. James

The tombstones tell this story.

Buried here are natives of Ballyvannin, County Clare, Donegal, Cork, Tipperary, and Wexford. And there are many more place names not quite legible.

Dust to dust. It comes to this, even for stone.

This is the cemetery at St. James at Sag Bridge Catholic Church near Lemont, a parish built in the early nineteenth century by immigrant laborers on the I&M Canal. This is where they worshipped, raised their families, and buried their dead.

"McMahan, Murry, O'Roark . . ."

Donna Slosowski, volunteer tour guide, measures her steps through the cemetery that surrounds the church. She points out the oldest markers, the most elaborate ones, and talks about the impressive structure at the center, the church itself. And like so many other things here, it is made of stone.

Slosowski is the unofficial parish historian. She brings 200 years of history to the conversation, conjuring up the memories of the people who founded this parish.

"They were the laborers who worked on the canal and in the quarry."

The labor force that dug canals throughout the East in the early 1800s was composed of immigrants, a great many of them Irish. Those laborers moved steadily westward, eventually working on the railroads, as the nation expanded and fulfilled its perceived manifest destiny.

But the availability of cheap land and the discovery of limestone near Lemont convinced many of them to settle here. The I&M Canal held great promise, and the quarry ensured that work would be steady. This was a parish built on that bedrock.

St. James parish, created in 1833, is the oldest in Cook County, one of the oldest in the state of Illinois. The descendants of some of its founding members are parishioners today.

" . . . Murphy and Sullivan . . ."

Two Irish families had donated several acres of high ground to the fledgling parish. Overlooking the Des Plaines River and the canal, the bluff was a fine place to build a church, everyone agreed. All they had to do was haul the rock up there.

There were probably dozens of men—sweating and toiling and driving teams of horses with wagons full of stone—who paused each day and wondered why they hadn't decided to build the church out of wood. It was a herculean task. After working six days cutting stone out of the quarry for a wage, these workers did not rest on the seventh, and it took them six years to get enough rock to the building site.

But the workers had an incentive, Slosowski says.

"The ones who got the most stones up here got the closest graves to the church."

It was an honor worth moving a mountain for. In those days, where one was buried told a great deal about who that person was in life. Surviving today, closest to the church, is the burial plot of the Jeremiah Day family, no doubt rock-haulers extraordinaire.

The lives of these people revolved around stone. It was the foundation of their church and its walls, too. Stone gave them life, and it would mark their graves as well.

"... *McGuinness, Coughlin, Byrne, Calhoun* ..."

The first stone church was a simple structure with a gabled roof. It was renovated and expanded around 1890, and the building has been relatively unchanged since the turn of the twentieth century. It and the cemetery are on the National Register of Historic Places.

"Come on," Slosowski says. "Let's go inside."

The interior is dim. A stained glass window behind the altar lets in a little light. There is a vague scent of incense, although none is burning. Slosowski settles into one of the polished wooden pews and points to the beams overhead.

"The woodwork is all original," she says softly. Even in a whisper, there is pride in her voice. She is a believer, and it's easy to see why: the strength of the beams, the glory of stained glass, the gleam of the gilded icons . . .

Slosowski's own background is fairly typical of the 750 families that make up the parish today. The largest ethnic group in the parish is Polish, a generation or two removed from immigration.

The Slosowskis emigrated from Poland around 1910, settling in the Back of the Yards district in Chicago, where there was plenty of work in the infamous stockyards that dumped waste into the Chicago River. Eventually the family moved west to the suburbs and found new homes in Willow Springs, just up the road from Lemont and St. James.

Slosowski started attending church here in the 1970s. She and her mother immersed themselves in parish activities from the start. Besides being the de facto parish historian and volunteer tour guide, she is a member of the Parish Council, the Ladies Guild, and the church choir. In a way, she's hauled a fair share of stone for St. James.

"... *Kelly, Ryan, Fitzgerald, Ford* ..."

She leads the way back outside. "We have a very active parishioner base," she says. "We've almost doubled in size in recent years."

Surrounded by history, her eyes are on the future. And strolling again among the dead, she talks about the living.

"I think initially people are drawn to the setting and the antiquity of the church—it's such a pretty place—but they stay because of the people," she says. "That's the way it was for me. The parish is very family-based. I know half the people at Mass."

*Donna Slosowski in
the St. James at Sag Bridge
Catholic Church, near
Lemont. Calumet–Sag
Channel Mile Marker
304.5, June 2002.*

Not far from where Slosowski walks is the McGuire family plot. It contains the nineteenth-century remains of Catherine and Peter McGuire and their four children. One died at ten months, another at seven years. Another lived a year and six months. The fourth made it to age twenty-five. Catherine and Peter outlived them all.

The inscriptions on the tombstones are reminders that the people who built the canal, who helped build the nation, were also building families. Or trying to.

"...O'Connell and Doolin, O'Brian, McHale..."

Some of the names have been forgotten, eroded into history. But even those faded, weathered tombstones speak volumes. They may barely whisper about who lies beneath, but what they say about us seems clear.

The stone will outlive all of us, standing for centuries in some cases, but it too will fade. It will yield to vandals and the elements until it reaches "the inevitable hour," as Thomas Gray calls it in his "Elegy," for even stone has a destiny in dust.

A Flower in Florence

Scrapbooks and Secrets

"All right," she says, pointing to the tape recorder, "now turn that thing off, and I'll tell you another story."

Katie Ballenger is the unofficial historian and keeper of the scrapbooks in Florence. In her living room, she'll march a visitor through a phalanx of photo albums and pages of brittle clippings, all chronicling the bittersweet story of a river town. But between the lines, left on the shelf and only whispered about, is a tale she doesn't tell to just anybody.

But if a visitor is patient, if he sips iced tea and looks through the books, if he hangs around and listens to the stories of bridges and floods and old times down at the dance hall, she just might open up and share that, too.

The town of Florence in Pike County was founded in 1836. It was originally known as Augusta, but it wasn't long before it adopted a new name, joining other towns in the valley—Rome, Naples, Brussels, Liverpool—that had linked themselves, at least nominally, to other parts of the world. It was indicative of the towns' lofty aspirations, perhaps, and in the nineteenth century, there was a lot of that going around.

A spring on the western bluffs of the valley attracted the first settlers. A wagon trail became a road, which attracted more settlers, and soon a ferry was crossing the river and steamboats were pulling in at the landing. The town grew. But unlike its namesake in Italy, the flower never fully bloomed in Florence, Illinois.

Its high-water mark probably came around 1850, when the Finley Hotel was built and storefronts lined the riverbank. It looked as if the bud was about to open. "At one time there were three or four churches

Florence resident Katie Ballenger on her front porch with Illinois River across the street in the background. River Mile Marker 55.5, September 2003.

Greg Crowther and his dog at the Spring Valley Access Area Dock, north side of river. River Mile Marker 218.4, March 2004.

The Station is a bit of a surprise. One wouldn't expect to find a jumping little joint in this otherwise sleepy town, but here it is. This is a blue-collar, pool-shootin' kind of a bar. There's a young crowd here, but there's classic rock on the jukebox—southern rock and country.

Bureau was founded in 1887 and named for a Frenchman, Pierre de Beuro, who had established a trading post here in the 1700s. It hasn't grown much since. A couple hundred people, at best, call it home. Situated on the outer curve of the Big Bend of the Illinois River, the town is also known as Bureau Junction, owing the latter part of its name to the railroads that merge here.

Under different circumstances, this might have been a big town. Not only did the rail lines cut through here, the Hennepin Canal connects to the Illinois River just south of town. This place had promise at one time.

The canal, originally called the Illinois & Mississippi Canal, was intended to be an extension of the I&M Canal. Goods could travel from Chicago to the Mississippi River at Rock Island in almost a straight line. It was the shortest route to the big river. But there were problems.

The idea of a canal to Rock Island was conceived in the 1830s, even before construction began on the I&M Canal. It was an ambitious project. It was to be about the same length as the I&M, but it would take more than twice as many locks, thirty-three. Plus, it would need nine aqueducts and feeder canals and man-made lakes to supply water. But the biggest hurdle was money. State financial problems delayed construction until the 1890s. By then it was too late.

More ambitious transportation ideas were already on the table. While construction was underway on the canal, the locks on the Illinois and Mississippi rivers were enlarged, allowing for bigger boats on those rivers. The new canal was simply too small. By the time it was completed in 1907, it was already obsolete. The Hennepin Canal was the stillborn dream.

Nicholas Malooley would gladly reveal more about the town and the Illinois River at this point if he knew more. So he calls to his best friend, "Hey, Greg! Come over here!"

And he confides, "He knows as much about the river as anyone around here."

Greg Crowther and Nicholas Malooley were born three days apart about twenty-five years ago, and they lived next door to each other growing up in Spring Valley, just a couple of miles up the road. They went in different directions after high school, but they're still close.

Malooley is a student, studying health and medicine; Crowther is a heavy equipment operator, a union man, driving dozers and cranes. These two young men have different dreams.

Crowther made a home right here. He married a schoolteacher and is about to become a dad. He just learned a few days ago that the baby will be a boy.

"My parents took me out on the boat as soon as I was old enough to walk," he says. "We'd go down the river and find a spot, and we'd just enjoy ourselves and whatnot. I'm sure my son will do the same. He'll be a river rat, no doubt."

Crowther inherited a slip down at the Spring Valley Boat Club from his father. Those boat slips are in demand, and it's not easy to get one. Greg, someday, will pass it on to his boy. Boating is an ingrained tradition in his family. And part of one's inheritance around these parts is a key down at the marina.

"It's nice to have roots."

As Crowther is talking, Malooley is pensively sipping his beer, listening as his buddy waxes philosophical about his family, his friends, and this place they all inhabit.

"People have a different attitude toward life on the river," Crowther says. "It's a natural community. Everyone is your friend on the water."

Malooley suddenly lifts his beer in a toast.

"Man," he says to Crowther, "I can honestly say that you have brought me to respect the river more than anyone I know."

It could be the beer talking, but the truth is in there somewhere. The guy who can't wait to get away from here has a soft spot for the "armpit" after all. Whatever happened to Hawaii and California?

"Oh, I'll probably end up back here," he admits, but he doesn't sound distressed about that prospect. He takes a ribbing for his earlier comments, and someone orders another round.

"And if I ever have kids?" he says, nodding in Crowther's direction, "I'd probably raise them here."

The water in the Hennepin Canal flows slower than the beer in the tap at the Station. But it still flows, not far from where these two young men sat talking about it.

The canal is a recreational site today; it has been since the 1930s. It is one long state park, a unique ribbon of land that starts here at the Illinois River and cuts clear across the state—not exactly its intended purpose, but things change. Like rivers. Like dreams.

Life on Meredosia Island

A *Survivor's Tale*

The decurrent false aster is native to the lower Illinois River Valley, and unlike most plants it relies on the flood for its survival. The seasonal swelling of the river strips off competing plants in the sandy, moist soil of the flood plain. The waters that submerge it only make it stronger.

But the river's flood cycle has been disrupted, and the once-abundant false aster has retreated to a few isolated stretches of the Illinois River, a threatened species.

When the conditions are right, though, its tiny white petals can be spotted here and there along a creek bank or at the edge of the timber.

Dorothy VanDeventer remembers clearly the day she met the man she would marry. It was November 27, 1936, the same day Governor Henry Horner was in town to dedicate the new river bridge.

There was a parade and an orchestra, and hundreds of people had traveled to Meredosia for the

Self-proclaimed "river rats" of Meredosia in the Meredosia Historical Society and River Museum. From left: Harvey Dean, 75; Larry Edlen, 59; Leonard Easley, 87; Darrell McDannold, 61; Sid Logsdon, 87; Earl Edlin, 81. Seated in front: Dorothy VanDeventer, 82. River Mile Marker 71.3, February 2001.

"You just don't let it rock you," she says. Simple as that. "You go on because life just has to go on. My husband was like that. He lost beautiful crops, just before harvest. He was hurting, but he didn't cry about it. There's no sense crying about it."

So Dorothy and Van found themselves packing up Rebecca and evacuating the island with greater frequency.

"People would say, 'It's terrible, why do you keep going back?' Well, it is terrible," she says with a little shrug. "But it was home."

After each flood, Dorothy and Van found themselves with fewer neighbors, most of whom sold their flood-prone land to the duck club. By the 1960s, the VanDeventers were almost the only year-round residents on the island.

When John C. Anderson, owner of the gun club, died in 1971, he donated his property to the Nature Conservancy, and the U.S. Fish and Wildlife Service took control of the land two years later. Today it is the Meredosia National Wildlife Refuge.

What was good for the goose wasn't necessarily good for the VanDeventers, and the 1970s were a time of uncertainty and hardship. Van worked for the Fish and Wildlife Service for a while, and the family was allowed to stay in the house. But the river wanted it also.

A torrential flood came in 1973, and other big floods followed. The house, although still standing, was battered. Eventually, they were forced to abandon it.

"You couldn't keep the river out," she says, growing quiet with the memory. "I've had a lot of changes in my life, but that was the hardest: moving."

But, still, you didn't cry about it.

Not long after they left the island and while they were preparing for their fiftieth wedding anniversary, Van died of a heart attack.

"I used to go and spend time up there, right after Van died," she says. She'd either drive by herself when it was dry, or someone with a boat would take her out and drop her off for the day. She'd take a lunch and walk around. "I'd just listen to the birds. It was home. I wanted to go home."

It would be easy to be bitter. She could blame the government, or the levee-builders, or the duck hunting clubs. She could blame the river itself. But she blames no one.

"You don't get much joy out of life if you do that," she says.

Whenever the road was open, she'd drive out to the old clubhouse. Sometimes she'd stop and walk through the place. All the memories were stored in that vacant and battered building.

She could hear the chatter and banter of hunters as she worked in the kitchen. She could still see that old stove flicker to life under her match and smell the savory aroma as she stirred the pot of duck chop suey. Van would come in the back door, kicking off his boots and slapping his coat on a hook.

She'd remember the quieter times, too. After the

baby had gone to sleep, she and Van would sit out on the porch to catch a cool breeze and share a dream.

Not long ago, she drove herself out to the island. It was a pretty, cloudless morning. As always, she drove out to the old place. Someone else was already there.

It was a refuge worker. He was on a tractor. He was tearing down the house and pushing it into a hole in the ground. The sight hit her like a punch to the gut.

Dorothy knew the man on the tractor—knew him enough to consider him a friend—but she didn't stop; she didn't wave. Eyes forward, she just kept driving, drove all the way down to the timber line.

There she got out of the truck and walked, straight and stoic. The birds flitted between the trees. She went down to that spot by the pond and sat on that log she once fell asleep on. She folded her hands in her lap. She gazed at nothing in particular, and she saw all of it.

It took a little while, but a tear finally formed. And she cried.

Sid and Earl

River Rats and Lifelong Partners

Between the two of them, Sid Logsdon and Earl Edlin have harvested almost 170 years of memories from the Illinois River. There would be more than a little overlap, of course, because Sid and Earl spent a lot of those years together.

They met as young men in 1939, and they've been hunting, fishing, and trapping together ever since. Both came from longtime river families, and a mutual respect and compatibility led to a partnership that has lasted most of their lives. When they married sisters—Mary Louise and Francis Bradbury—the two friends became family.

"We've been together our whole lives," Earl says with a gravelly voice that sounds like a friendly growl. "And we could talk your ears off."

Sid nods. "We could tell you stories all day and never repeat a one." Six years older than Earl, Sid has a higher, more inflected voice that he's honed at the barn. He still calls a square dance from time to time.

The two self-proclaimed "river rats" have come to the river museum in Meredosia today to talk about the old times and their lives on the river. It doesn't take much to coax a tale out of them.

They talk about steamboats—"you can tell them apart by their whistles"—and kerosene lighthouses along the river channel. They talk about mussels and shelling—"used to be you could make a living off this river"—and fishing. And there are hunting stories, too, like the time they cornered a mink in a groundhog hole.

"He's a little better shot than I am with a shotgun," Sid says of Earl, "so I got me a grapevine, and I said, 'I'll chase him out and you shoot him.' So I ran the grapevine down there and gave it a shake, and out it comes—*whoop!* And off it goes through the brush as

Looking west from Indian Creek, north of Meredosia. A lead barge heads downstream. River Mile Marker 78.8, September 2004.

fast as it can, and it charges right at that old 12-gauge and—*boom!* He missed it! But that mink turned right around and came back at me, and I caught him with my own two hands."

Sid breaks into a laugh, which might make a listener think that he's making the tale up, but Earl is sitting there nodding.

"Yup," he growls, that's what happened, all right.

There's more to these guys, though, than a good yarn and a laugh. Each is a serious practitioner of the folk arts.

Earl, a recipient of Illinois Arts Council grants, has participated in the state's folk apprentice program, passing on his knowledge to a younger generation. His expertise is building wooden fishing baskets, a handmade cylindrical contraption that operates on the same principle as a hoop net.

Sid has received a Smithsonian Institution grant to pass on his knowledge of building cedar boats. He traveled to the nation's capital in 1997 to demonstrate his craft. Closer to home, he's conducted public workshops on the now-dead mussel industry, showing how crowfoot hooks are made and used.

As Sid puts it, "There's no equipment that's used on the Illinois River that we can't build, the two of us. Nets, baskets, boats, hooks . . ."

Like a lot of the old-timers, Sid and Earl know their way of life is passing away. Through their efforts—from exhibitions in Washington, D.C., to demonstrations in museums in Illinois—the two are helping to hold the head of a dying culture above the rising waters of time.

To that end, there's nothing they like to do better than tell tales.

"Earl saved our daddy-in-law from drowning once," Sid says. He tells how he and Earl would take their father-in-law out shelling because the older man didn't have a boat with a motor. One day old man Bradbury fell into the river, and Earl grabbed hold of him and got him back into the boat.

"Yup," Earl says. That's what happened.

Sitting side by side, Earl Edlin and Sid Logsdon look like old buddies, but they are at the core different animals. Sid has a mischievous streak in him and a boyish charm. He's popular at the dance barn. He wears a constant smile, and he's never far away from a rip-roaring laugh.

Earl is more serious. He doesn't keep his opinions tucked away; they're out there on his sleeve. He's point-blank. "There's going to be a day a'comin'—and it ain't too far off—that there ain't gonna be a fish in that river." And, "The government has screwed this river up from day one."

The two share the memories and the opinions—after all these years, they're still partners in that regard—but their styles are different. While Earl's up on the soapbox, Sid's weaving a tale.

"I never had any use for a game warden," Earl says, and Sid conjures up a memory of the time they caught a few fish that were undersized. "So we dressed them and put them in the game warden's car."

That gets Earl to nodding. "Yup."

The day's getting late now. It's time to head home.

Sid's on his feet saying good-bye. He was right about one thing: they talked all day and never told the same story twice. Still light on his feet, Sid steps into the doorway.

"I'll see ya, Sid," Earl says, and there is a solemn tone in that scratchy voice. Sid flips a little wave and leaves with another grin. Then he closes the door behind him.

As Earl prepares to leave also, the words he'd spoken earlier resonate clearly now. "Pretty soon," he said, "there won't be any of us old-timers around anymore. We're losing another old friend every year."

There is plenty of life left in these two octogenarians, and although their way of life will pass—that can't be stopped—their efforts to preserve the old crafts will help keep the memory of that culture alive.

For now, though, Earl is going home. He says good-bye. And then he, too, slips through the doorway, and he's gone.

Buck Barry of Hardin

Building a River

Buck Barry is standing at his door, a pint of raspberries in his hand. He's holding the screen door open with his other hand, and he's pondering the question.

"The Illinois River? Well," he says, "I helped build it."

And with that, he extends an invitation into his home and into his life.

Aloys "Buck" Barry lives atop the dividing ridge in Calhoun County, the sliver of land that separates the Illinois River from the Mississippi River. He's well into his nineties—"a 1909 model," he says.

Age has started to slow him some, but every day he still drives his 1989 Dodge Aries into Hardin, where he's a regular at Royce's Restaurant and the Barefoot Restaurant and Bar.

In a manner of speaking, Barry did help build the Illinois River. He worked with the U.S. Army Corps of Engineers in the 1930s, during creation of the Illinois Waterway, when the navigation channel was enlarged and the current locks and dams were installed. And in the 1920s, he was on the gang that built the bridge in Hardin, the first—and only—river bridge into Calhoun County.

Buck Barry is a living link to the days of the steamboats, before the great changes, to the days he could "swim across the river and think nothing of it."

"I was like a rat. I was in the river all the time."

It was a different animal, though, that lent him his name. While growing up as one of eight children of James and Anna Barry, he and his sister Maureen were inseparable, like the team of mules at the Barry place, Buck and Jack. His nickname stuck. Thankfully, Maureen's didn't.

Today, the memories sometimes flicker by like in a nickelodeon, moving pictures fractured over time. And then there is focus, a time will crystallize, and a story will emerge.

Buck Barry sits alongside the western side of the Illinois River, upstream of the Joe Page Bridge at Hardin. River Mile Marker 21.8, September 2003.

"My dad ran the boat landing down there, best landing on the river," he says. "He shipped apples to Peoria and St. Louis, thousands of barrels of apples. You can't imagine how big those steamboats were, those big paddlewheels. There was the *Piasa* and the *Golden Eagle* . . ."

The Barrys owned 109 acres at the water's edge on the southern end of Hardin. The Jersey County Grain Company elevator sits on that land now. Until his death in 1929, James Barry shipped produce and off-loaded goods for the town at their landing and operated a cider mill that churned out thousands of gallons of cider during the season.

"There'd be an acre of barrels sitting out there, just fermenting," Buck says, grinning at the memory. "And there'd be a spigot in the top of the barrels, see? And you'd have to turn that spigot to let the gas out, or it would blow up. Well, the kids would get on up there with a straw or a reed, and they'd stick it down the hole and suck the cider through the straw. Oh yeah, I was as guilty as any of them."

Barry talks of the old days as if they were yesterday, and he's old enough to have heard the tales of the nineteenth century firsthand. He tells the story of how his grandfather, Peter C. Barry, first came to Calhoun County. Dr. Barry had just finished medical training in St. Louis and was on his way to Peoria to start a practice there when his boat landed at Hardin.

"There was a lot of sickness in the county—I don't know if it was the flu, might have been—and some-one knew he was on that boat, so they asked if he'd get off there and help out some," he says. "He never made it to Peoria."

Dr. Barry built a home, an imposing structure that still stands on the bluff south of town, and married Jersey Smith. He founded a newspaper, the *Calhoun Herald*, and somehow along the way they raised half a dozen children, the oldest of which was James, Buck's father.

Calhoun County is bordered on the north by Pike County and by rivers everywhere else. At the turn of the last century, what few roads came in from the north were not much more than ruts and trails. The county's peninsular shape and hilly terrain discour-aged the railroads, which serviced East Hardin across the river. Not a mile of rail has ever been laid in Calhoun County.

The river also was a purveyor of entertainment to the community, and Barry remembers those days well. "I've danced a lot in my time. That's all we had," he says.

He recalls that he and his buddies—this would have been in the late 1920s and early 1930s—would drive a Model T down to Brussels because the Wittmond Hotel would ferry in a big band from across the river.

"It was twenty miles away and would take at least an hour to drive there, depending on how many flats you had."

A better time was had on the riverboat.

"They used to run an excursion boat, the *Idyllwild*, and there'd be dancing on the boat all the way up to Kampsville," he says. "Old man Clark owned the dance hall just north of the ferry, right on the beach."

About this time, Buck met Mildred Surgeon, fell in love, and got married.

Things started happening all at once for him. He started a family at about the time he started a new job. He went to work for the Army Corps of Engineers, making more money than he'd ever seen before.

"I went from a dollar a day to forty cents an hour."

The corps was charged with replacing the old locks and dams and finishing the new Illinois Waterway. He worked on a boat, helping survey the river and lay out the channel. He and his crew would take soundings, working with a dredging outfit. Sometimes the work was dangerous, and one time it almost cost him his life.

"The old dam had washed out, see?" he says, telling the story about how they were working up at Kampsville. "I had a sounding pole, and we were taking readings every twenty feet. Well, that current caught the pole and me, and over I went."

The water pulled him into the opening in the dam and washed him right on through. Another boat downstream fished him out of the water. It was a close call but one that taught him a lesson: You respect the river at all times.

Barry also piloted an inspection boat for the corps, a 250-horsepower Cris-Craft that could run thirty miles an hour or better. He'd ferry corps officers from Chicago to St. Louis in that boat. That's what he was doing when Pearl Harbor was bombed. He was thirty-three years old and a new father when he was drafted and sent to the Pacific, where he served as an engineer on a picket boat.

After the war, he came home to Calhoun County, and he and Mildred had two more boys. He got into farming. "We raised a lot of damn apples."

Sometimes Barry's memory is so clear he can put his finger on facts as if he'd just read them. While telling stories that are half a century old, he can rattle off names of people and recall their hometowns without pausing to think. There's two war buddies, Louie Miller from Denison, Iowa, and Bill Hirsch from New York; and there's Bill Nolan from Meredosia, who was a lighthouse man on the Mississippi River back in the 1930s. There seems to be an endless list.

A good memory, though, can be a burden, too.

"I'm a lucky devil to live so long," he says. But his smile slowly fades as he recalls a name from some older time or as a familiar face from the past flickers for a moment and then dissolves. "All my buddies are gone. That's the bad part."

Mildred died in 1997 on December 31. He gets quiet when asked about her. She was a few weeks shy of her ninetieth birthday. And his sister Maureen, the other part of that mule team? She passed away just this last spring.

Yes, that's the problem with people. They tend to drift away like apple blossoms and oak leaves, like sticks floating off on the current.

Looking west from the City of Chicago Police Marine Unit Pier to Lake Shore Drive and the Chicago skyline. Chicago River Mile Marker 326.5, April 2001.

Beginnings, Perhaps

At the Lake in Chicago

Depending on one's perspective, the Illinois Waterway either begins or ends in Chicago.

At the beginning of the twentieth century, the Chicago Sanitary and Ship Canal effectively reversed the flow of the Chicago River, and the Calumet–Sag Channel did the same for the Calumet and Little Calumet rivers. Courtesy of human engineering, the mouths of rivers became their headwaters.

Looking east at the beginning of the Calumet River, toward Lake Michigan. The Elgin, Joliet and Eastern Railroad lift bridge is the first span across the river when entering from Lake Michigan. The South Ewing Avenue Bridge near 92nd Street can be seen at the right. This is the Tenth Ward. Calumet–Sag Channel Mile Marker 332.7, March 2000.

*Second mate Toby Baker on the bow of a tanker
barge, in radio contact with Captain Tom
Flowers of the* Orleanian *as he enters Starved
Rock Lock and Dam. River Mile Marker 231,
October 2000.*

CHAPTER TWO   An Open Channel

It's 3 A.M., silent but for the bugs chattering and buzzing in the blazing lights overhead. Concrete and pipe, water and machine. This is the lock and dam at Dresden Island. It's quiet at this hour, even as 22,000 tons of steel and cargo glide in from out of the darkness.

The first mate talks into his radio in a slow, low monotone. "Three-and-a-half wide, closing easy, about fifty feet to the gate." He's out there on the lead barge, talking to the pilot in the wheelhouse.

The pilot guides the tow in, gently coaxing the controls with his fingertips. Down below, the two V-12 diesel engines are humming softly, an awesome power held in reserve. That power will be needed sometime, but not now. This maneuver calls for finesse.

"Three wide, making a slow walk in, about thirty feet to the gate, two-and-a-half wide and closing."

The entire tow—fifteen barges and a towboat—is too long to fit into the lock. It will have to be broken down and sent through in two pieces. The towboat will push the first group of barges into the lock, deckhands will cut it free, and the boat will back out, leaving behind two deckhands and 600 feet of barges in the chamber.

Because the first set of barges, called the "break,"

is powerless, a mechanical "mule" anchored to the lock wall will winch it out of the lock when the time comes. That's when it gets dangerous.

"Almost abreast of the pin, a foot to the wall and closing. Steady."

Steady is the operative word. A deckhand cannot afford to take his mind off the task. When the mule pulls the break out of the lock, the barges are free-floating with nothing to stop them except ropes and the barge cowboys on the deck.

A deckhand tosses a line three inches thick and lassos a pin on the lock wall, then loops it loosely around a kevel on the barge. He allows enough slack in the rope so he can check the speed of the break and eventually halt it. There are thousands of tons riding on the end of his rope. If his line is too slack or too tight—if it gets hung up on anything—it could snap or slingshot anywhere. A seasoned deckhand listens to the rope. He trusts his ears.

The rope tightens, *crick, crick!* The deckhand feeds it a little more slack and then loops the kevel again. And again. The rope creaks, a fat, taut piano string. The pitch gets higher and higher, and it sings out one final *CRICK!* . . . and falls silent.

"We're at the pin," he tells the pilot. "All clear."

Looking east across the river near the remains of the nineteenth-century lock at Copperas Creek near Banner. River Mile Marker 136.8, November 2001.

This interaction between man, machine, and river is continual on the Illinois. The locks are in action all day long, every day of the year. Barge traffic on the river halts only for flood and fog and occasionally ice. Sand and gravel; corn, coal, and petroleum product; chem, steel, and scrap—so much relies on the navigability of the river.

If the river were the circulatory system, the locks and dams would be the heart. Steel gates weighing more than 250 tons each open and close like massive ventricle valves, and thousands of tons of coal, stone, and chemicals are pumped out every day. Every year, more than 700 million bushels of corn flow through the system.

One can take the pulse of the economy by counting the grain barges and doing the math.

There are nine locks on the Illinois Waterway, including the Thomas J. O'Brien Lock on the Calumet and another at the mouth of the Chicago River.

The elevation of Lake Michigan is about 160 feet higher than the level of the Mississippi River, and the locks and dams gradually drop the water level along the length of the waterway, about 330 miles. The biggest drop is at Lockport, the first lock out of Chicago, where the level changes almost forty feet. Another significant feature of this lock and dam is that it regulates the water level all the way back through Chicago, and consequently the discharge at

A local fleet boat moves upstream to shift the order and makeup of a fifteen-barge tow near La Salle. River Mile Marker 224, October 2000.

Looking upstream and east at the
Bartonville Petroleum Terminal Dock.
The midday, midsummer haze was a result of
high temperatures, which topped 100 degrees by
early afternoon. River Mile Marker 155,
July 1999.

this dam is controlled by the Metropolitan Water Reclamation District of Greater Chicago. The U.S. Army Corps of Engineers controls the dams downstream.

Lockport dam's flow affects the pool immediately downstream, which is controlled at Dresden Island Lock and Dam. The lockmaster there in turn adjusts the discharge at his dam to maintain normal pool level, and so it goes all the way downstream. It's the domino effect, and it starts at Lockport.

Those who rely on this system make up a diverse customer base, from the huge barge companies to the recreational canoeist. There are fishermen, speed-boaters, single-barge haulers, and sailboaters. The waterway puts tens of thousands of people to work, directly and indirectly, and it is vital to the health of industries and towns. It is the primary grain chute for Illinois farmers, and in times of war, the waterway earned its stripes.

This river was built, quite literally. And it was built to move product.

DEEP ENOUGH

Native Americans had long used the river to trans-port trade goods in dugouts and canoes, but what the Europeans—and later the Americans—brought to the frontier was the concept of engineering a more efficient river. The Illinois & Michigan Canal was just the beginning.

In its natural state, the Illinois River was a slow, shallow stream whose navigability was unreliable.

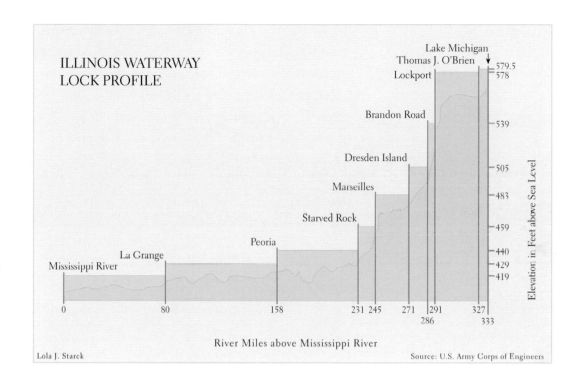

Boats moving upstream in the early nineteenth century often had to transfer cargo to smaller craft, and settlements sprouted up where these transfers were made. As the population grew, the need for an unimpeded river route became more important. By the late 1820s, shallow-draft steamboats were plying the waters of the lower river, and Peoria was already a busy port.

By 1848, the I&M Canal had opened a clear channel between La Salle and Chicago, solving the problem of shallow water and rapids, most notably at Starved Rock and Marseilles. At La Salle, cargo from packet steamers and sternwheelers would transfer to canal boats, which would be hauled by mules along

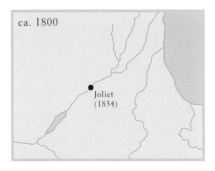

ca. 1800

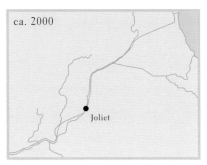

ca. 2000

the towpaths all the way to Chicago. Sometimes the steamboats increased their hauling capacity by pushing canal boats up and down the Illinois.

But navigability below La Salle was not a guarantee. Shifting sandbars and shoals were constant hazards, and when water levels dropped, the river was simply impassable. Several times during the 1850s, severe drought turned the Illinois into a trickle. Riverboats could not get beyond Peoria, and trade at the terminus of the I&M Canal was cut off for months at a time. There was a lot of money floating on the navigability of the Illinois River. And there were other issues, too.

The Civil War demonstrated the military importance of reliable inland waterways. A nation threatened by foreign powers and at war with itself could not afford to bottle up its troops, waiting for the creeks to rise. Quickly moving men and matériel from the Great Lakes to the Mississippi River was considered essential to national defense.

So after the Civil War, the federal government took a keen interest in Illinois's efforts to improve the waterway. And when the state's financial well went dry, the U.S. Army Corps of Engineers stepped in.

The project involved dredging and building locks and dams, the first of which was completed at Henry in 1871. A second lock and dam was built soon after at Copperas Creek, south of Peoria. By the end of the century, more locks and dams were added further downstream—at La Grange and Kampsville—and together the four dams created a series of navigation pools, ensuring a fairly reliable channel from the

Mississippi River to the western terminus of the I&M Canal.

Chicago, meanwhile, was grappling with a deadly problem whose solution would forever change the landscape of the river valley. And that solution would inadvertently carry the waterway into a new century.

Chicago's problem was its sewage and waste. To keep from polluting its drinking water, which was drawn from Lake Michigan, the city engineered a solution that would reverse the flow of the Chicago River and send its waste downstream via canal. After attempts to use the I&M Canal for this purpose failed, the Chicago Sanitary Canal was dug. It opened in 1900. Unlike the I&M, navigation on the sanitary canal, initially, was not a priority. It wasn't used for shipping until 1906, and its name was changed to the Chicago Sanitary and Ship Canal to reflect its new role.

The success of the canal and a variety of other factors—the demand for more water from Lake Michigan, increased pressure on the waste disposal system, the need for a shorter route to the growing port area south of the city—led to the construction of a similar canal, the Calumet–Sag Channel, which was completed in 1922.

The canals created a sort of hourglass transportation system. There already was a relatively free and clear channel from the Mississippi River to La Salle and now a wide, straight canal between Lake Michigan and Lockport, where the Sanitary and Ship Canal emptied into the Des Plaines River. The

Moving upstream and north near Chillicothe, a towboat with barges navigates the narrow, ice-choked channel during the coldest Illinois December since 1895. Temperatures for the month would average thirteen degrees below normal by the end of the month. Navigation was temporarily suspended during this cold snap because of the cold and the ice, which is a rarity on the Illinois. River Mile Marker 180.4, Christmas Day 2000.

bottleneck lay between Lockport and La Salle, where the river, despite the higher water level, was too unreliable.

The old I&M Canal, which was still used for some commercial purposes, was too small for the larger boats of the day. The canal that had played such a pivotal role in the development of the Illinois interior had become outdated.

Late in the 1920s, work on a new waterway began in earnest. The idea was to create a nine-feet-deep navigational channel from Lake Michigan to the Mississippi River utilizing the Chicago Sanitary and Ship Canal, the Cal–Sag Channel, and a new set of locks and dams on the Illinois and Des Plaines rivers.

It was a massive public works project involving state and federal resources. Locks and dams were installed at Starved Rock, Marseilles, Dresden Island near Morris, and Joliet. Because the water level of the river was to rise, levees and new steel bridges were built. To keep the city of Joliet from flooding, a monolithic seawall was constructed to contain the waterway as if it were an aqueduct.

The channel in Joliet is higher than the rooftops of some of the downtown buildings. Today as one travels the waterway, one can look down at the streets below, at Joliet's city center, at public housing units and Harrah's Casino.

Construction of the seawalls, bridges, and locks and dams took years. And then on June 22, 1933, while Chicago was celebrating its 100th birthday with the Century of Progress International Exposition, a gang of barges from New Orleans was pushed into

Chicago, marking the official opening of the Illinois Waterway.

The Illinois Waterway has been tweaked over the years. Retractable wicket dams were installed at La Grange and Peoria from 1935 to 1938, replacing the original dams there. And locks were added at the mouth of the Chicago River and on the Calumet. But today the waterway basically relies on the infrastructure that was poured in the early 1930s.

IN TIMES OF WAR

Both the state and the federal governments had a vested interest in the navigability of the Illinois River—the state from a commercial standpoint and the federal government from a military one.

As far back as the War of 1812, the possible military applications of the river were apparent. The British, antagonists of the United States until the latter part of the nineteenth century, had access to the Great Lakes through Canada. Access to Lake Michigan from the south could prove strategically vital to the new nation.

During the Civil War, rivers quickly conveyed gunboats, supplies, and soldiers. Indeed, thousands of troops shipped out on the Illinois River from places like Peoria, Havana, Beardstown, and Florence.

World War I pushed the United States into the role of world power, and during that conflict it became apparent that a reliable waterway through Illinois would have helped move troops and matériel from the nation's interior. Military interests had joined

Looking northeast at the Elgin, Joliet and Eastern Railroad drawbridge near Morris and just downstream from Dresden Island Lock and Dam. The lightly used bridge presents the narrowest opening for navigation on the waterway other than on the Chicago River. It has a horizontal clearance of 120.5 feet. (See also page 182.) River Mile Marker 270.6, January 2001.

commercial interests in the push for navigational improvements. It was an imperative that fueled the drive to build the waterway.

Eight years after the waterway's opening, the United States was drawn into World War II, and the waterway was immediately pressed into duty. Submarines built at boat works in Wisconsin were floated through Chicago and down the river, bound for the Gulf and the Pacific. Explosives were manufactured at the Joliet Arsenal and sent out on the waterway. But perhaps the river's greatest contribution to the war effort came from a shipyard in Seneca.

From 1942 to 1945, the Chicago Bridge and Iron Co. turned out 157 landing ship tanks, or LSTs, at the Seneca shipyard. The troop and matériel carriers were built at river's edge and floated out before being thrown into action in both theaters of the war. At its peak, the shipyard employed 11,000 people.

It's unlikely the Illinois Waterway ever again will play such a prominent role in times of international conflict, but during World War II the river answered the call.

MOVING PRODUCT

Because moving heavy material by water is efficient, settlements and industries have always clustered near the river. And because a natural and fairly reliable channel reached it, Peoria early on became the largest city on the Illinois River. More than seventy-five steamboats were docking at its wharfs in 1858.

Today, grain and coal are moved from various places along the waterway, from Hardin to Joliet. Power plants, chemical processing facilities, and oil refineries line the banks from Morris to Romeoville. The Cal–Sag Channel in Chicago is one long industrial corridor. A brand-new asphalt plant in Lemont receives aggregate by barge every day, and every day it and similar plants send out product on the waterway.

"This river is absolutely vital to our operation," said John Gobert, a spokesman for Citgo Refinery near Lemont. Although most of Citgo's product, primarily diesel and gasoline, is pushed through pipelines, 30 percent of its production is shipped out by river.

Without the river, that product would have to be trucked out, piped out, and moved by rail, potentially more expensive options. While short runs and smaller shipments might be better served by rail or truck, the barge industry's advantage is its ability to move bulk. A single barge can carry as much as fifteen railroad tank cars. A fifteen-barge tow can haul as much as 870 semi-tractor truck trailers.

Citgo's story is echoed up and down this stretch of the river. Tank farms and power plants are strung out along the waterway, pipelines and high voltage lines crossing each other in Mondrian patterns. Some power generation facilities rely on coal, which is delivered by rail and barge, and river water is used for cooling and steam. There's a lot of energy flowing from this grid.

Not all the businesses here are goliaths like Exxon-Mobil and Commonwealth Edison, and not all the barge traffic is handled by the colossal American Commercial Barge Lines, the largest river carrier in the world. Smaller towing companies perform vital intermediary roles on the waterway.

At Lemont, the terminus for heavy barge traffic coming upriver, Hannah Marine and Garvey Marine both have harbors. Both companies provide fleet service with their smaller boats, helping dismantle large tows and sometimes taking a single barge or a smaller string of barges farther up the canals. The waterway narrows at Lemont, and the Chicago Sanitary and Ship Canal and the Cal–Sag Channel are both too tight for a fifteen-barge tow.

Garvey also operates harbors in Morris, Seneca, Ottawa, and Pekin. Hannah Marine has a bigger operation on the Great Lakes and, unlike Garvey, owns its fleet of barges. There are other fleeting services up and down the river.

According to some estimates, more than 350 businesses have terminals on the Illinois Waterway. And each year, 44 million tons of product move, half of which is agricultural.

Downstream, almost every town has a grain elevator. Because a single barge can haul 52,000 bushels of grain and a full fifteen-barge tow can hold 22,500 tons of corn, elevator companies on the river have cheaper transportation costs than do their inland competitors. Those savings translate into better prices paid to farmers.

In Morris, three agribusiness giants—Archer Daniels Midland, Louis Dreyfus Corporation, and Cargill-Continental—have facilities, and farmers as far away as Kane County and De Kalb regularly haul their grain there.

"It takes us an hour to drive to Morris," says Herb Ruh, a longtime farmer from Kane County. "And we pay that cost." But the price he gets for his grain at the Morris elevators makes the drive worth it.

Home at the Helm

Somewhere on the Illinois

Every morning, they wake up in a different place. The scenery changed overnight, and time rolled on. They are in a different world.

On the boat, crewmen can easily lose their sense of place. On land, they are centered. They know exactly where home is. On land, they go to sleep confident that when they wake, their beds will be in the same spot they left them. It takes awhile for a person to adjust to life on the river.

For the past month, Tom Flowers has been home, guiding his Allis-Chalmers tractor through the fields of his small farm just south of Beardstown. And he has been sharing moments with his family, moments that most people take for granted.

He watched his daughter, Sarah, twirl around in the yard until she got dizzy and fell into a bundle of giggles. He watched Sam, his eleven-year-old son, run headlong across the field. And in the evening, he sat with his wife while the sun went down. He and Cathy talked into the night, and sometimes they'd laugh. For thirty days he's been able to hold her.

On the telephone and checking incoming shipping data on his computer, Captain Tom Flowers in the wheelhouse of the Orleanian *guides a full load of barges through a morning fog near Banner Marsh. River Mile Marker 143, August 2000.*

But for the next month, starting today, Tom Flowers will do none of that. His home will be the boat. His place is the wheelhouse. And every morning, he wakes up in a different world.

Flowers boards the *Orleanian* as it's waiting to lock through at La Grange. He's the captain, the man in charge, back for his regular tour of duty. The *Orleanian* is registered out of Paducah, part of the fleet of boats crewed by Western Kentucky

Navigation, a subsidiary of American Commercial Barge Lines. It is one of the cleanest and largest boats on the Illinois.

Members of his crew greet him with no more formality than a smile, a handshake, and a "Howdy, Tom." He's respected and well-liked. The liking part is okay, but it's the respect part that's vital. If a captain lacks that, there's bound to be trouble with the crew.

The pilot who had been filling in for him says a

few words—nothing more—and tosses his duffle into the johnboat. The baton has been passed. He's off to shore, off for thirty days now, and there's reason to smile.

Flowers stows his gear in his cabin and heads to the pilothouse. It's late afternoon, and he dives into work, checking reports, reading the log, flipping pages, and clicking through screen after screen of computer data, reams of printouts and electronic bytes. Every now and then, the radio coughs up a voice. Welcome back to work, Captain Tom.

He's been doing this for about fifteen years, and before that he did harbor work for a fleet service and labored out on the barges as a deckhand. He knows the routine, and he gets down to business. He needs to know what he's hauling and where it's going. And pickups? He checks his orders and reads the river condition reports. What's the weather forecast? Any problems with the crew?

From the reports, Flowers learns what he's pushing—salt to La Salle, molasses to Chicago and a Great Lakes transfer, styrene to the refineries near Romeoville. He is to drop the tow in Lemont and "make the turn" there, meaning he is to pick up another string of barges and head back down. The river's at normal pool, and the weather forecast is good. It should be an easy run.

The best report he gets, though, is that the crew is a good one: there's a competent first mate and no troublemakers among the deckhands. "All it takes is one bad apple in the crew, and this could be a real long run," he says.

The horn at La Grange sounds, and the radio comes alive. Bring it in, Captain Tom, and Flowers nudges the tow, fifteen barges, toward the lock.

On the boat, almost everyone works a six-hour shift, six on, six off, around the clock. The captain is spelled by the pilot, the engineer by the assistant engineer, and the deckhands work in two two-man crews. The fifth hand is on call, pulled into action whenever there's work to do, no matter what time it is. It's always the green deckhand who gets to be the fifth hand.

The engineer and his assistant are among the best on the waterway in Tom's opinion, and the pilot who will be swapping shifts with him is an old buddy and a good man. Perhaps the most important member of the crew, though, is the cook. Charged with providing three meals a day seven days a week, the cook can make the difference between a good run and a bad run. Cooks work weeks on and weeks off like other crew members, and not always are they assigned permanently to a particular boat.

"I've been on some runs where everyone was grumbling because the food was bad," Flowers says. Well-fed bodies mean happy crews.

The *Orleanian* is what is known as a St. Louis class boat. The galley, or kitchen and eating area, is in the center of the boat, and the table is round. It's a symbolic nod to the egalitarian nature of the outfit. The captain eats with the crew as an equal, not at the head of some elongated piece of furniture. "We're all equals here," Flowers says. "We just have different jobs to do."

In that spirit, all crew members have access to the pilothouse, even the deckhands.

Once clear of the lock, the *Orleanian* begins the push upstream. Night is falling fast. Flowers flips on the light and tests the high-powered beam. It pays to keep an eye on the shore, to know where you are, so you know where you're going and remember where you came from.

Down on the deck, the silhouette of the second mate moves across a barge in that space between illumination and darkness. Flowers watches him, lost in thought.

A deckhand's pay starts at eighty-five dollars a day. He works thirty days and is off fifteen. He's expected to stay clean and sober; there's no alcohol on the boat. And he's subjected to random drug tests.

It's hard work, and it's a tough routine. Having a family life is difficult for most, impossible for some. But if a man needs to dry out or get away from some ground-bound trouble, if he needs to sock away some cash and avoid temptation, if he wants to see a piece of the country, then working the barges isn't a bad job.

Some deckhands are pretty rough; many have police records. Some are just kids without a past wandering across the present without a clue about the future. Some make a career out of it, as Flowers did, but a good deckhand is hard to keep.

Luke Moore, vice president of Western Kentucky Navigation, says he has to hire ten deckhands just to keep two. Sometimes Moore will find one who wants to work his way up. He might aspire to be an engineer, tuning motors and greasing gear boxes in the belly of the boat. Or he might want to become a pilot. More often than not, though, the deckhand is transient. He won't be around long.

As commander of the vessel, Flowers cracks no whip, but he doesn't fraternize with the crew much, either. Still, every now and then, one of the deckhands will come up to talk with him. At forty-two, Flowers is too young to be a father figure to these guys, but he's accessible. And he's a nice guy. It's easy to unload on him.

"Sometimes it's just petty stuff, complaints about another crew member, like so-and-so isn't doing this or that," he says. "And sometimes it's about trouble at home. You know a guy's work is no good if he's worrying about his girlfriend or his wife back at home."

Flowers is a quiet man and doesn't offer much advice. Mostly he just listens, he says. And sometimes he hears too much.

"It floors me, some of these younger guys, how they talk," Flowers says. "They're up here talking on the phone, and I hear them say, 'Did you pay this?' or 'Didn't you sign that?' and then they'll say, 'Well, goddamn it, bitch,' and something else, and then they say, 'Good-bye, Mom.' They were talking to their own mother."

At this stage of his life, Flowers seems far removed from the deckhands, but he has more in common with them than they might realize. Long before the boats, he was a wild, roughhousing man who loved motorcycles and drinking and good times. Drive fast, take chances. It almost killed him.

He was on his bike heading home from the tavern one early morning when he fell asleep going sixty miles per hour. He woke up in time to see the grill of a semi coming at his face.

He survived the wreck, and he was smart enough to realize that there's more to life than bars, bikes, and babes. He got serious about working. He met Cathy. They settled down on a piece of ground and raised a family. That's when Tom Flowers found out what's important to Tom Flowers: family, solid ground. He found out where home was.

A deckhand might think the captain has an easy job on the boat. The window of the pilothouse works both ways. They'll look up from the steel decks, where they're lugging around chains and slamming hammers, and they'll see the fresh-faced captain. He's not even breaking a sweat. But piloting this rig is no promenade along the lock wall.

The responsibility is huge. He's pushing 23,000 tons of cargo in fifteen barges that stretch out a quarter mile—millions of dollars of liability. So many things can go wrong and at any time. A line could break, a deckhand could fall, the boat could run aground. He could lose sleep over so many things, but he can't afford to because he needs his rest.

"Mentally, this job can wear you out," Flowers says. "You have to stay alert all the time, or you can get into trouble in a hurry."

The biggest worry is about the lives at stake. Death is a very real thing on the river. Flowers knows that all too well.

It happened several years ago, on the Fourth of July. He was piloting a tow through Peoria Lake. It could be tricky water at any time of the year because of the expanse of shallow water and the narrow channel. The water was up a little, which might have been a blessing at any other time, but good weather, high water, and a holiday had attracted a lot of boaters, some of them inexperienced.

There were sailboats cutting into the wind, fishermen in johnboats slipping in to get to the spawning beds, and pleasure craft darting in and out. It was a cauldron of chaos.

Flowers was pushing fourteen barges, so the tow was notched in front, meaning the barges were three abreast for four rows and then two more barges were in the front. He can still see the little speedboat coming up on his side, trying to pass in front of the barges.

"He cut into that notch," Flowers says. "I guess he thought he'd cleared the whole tow, but he hadn't. He just got up in there and couldn't get out. I couldn't do anything about it. We went right over him."

A man had handed control of the speedboat to his teenage son, who was at the helm when it was swamped and overwhelmed by the barges. They both drowned. Although he was cleared of any liability in

Photographed from an upper deck, the towboat's towknees are centered behind a full pack of barges being pushed upstream. The captain is guiding the approximately 105-feet-wide load through the 307 feet of available horizontal clearance of the Peoria and Pekin Union Railroad drawbridge on the south side of Peoria. The railroad leaves Peoria to the southeast and, once across the river, runs the border between East Peoria and Wesley Slough. River Mile Marker 160.7, August 2001.

the accident, the incident almost forced Flowers from the pilothouse.

"I cried," he says. "It tore me up." The incident has haunted him for years, and it's as if the river won't allow him to forget. "For three years in a row on the Fourth of July, I'd find myself almost in that exact spot on Peoria Lake."

What eats him is the knowledge that there are two people who will never go home again. They'll never be able to share dinner with the family, to run across a field and fall down laughing. They'll never be able to kiss their children good-night. And they'll never be able to take any of that for granted.

Nighttime on the river is a silent black shroud, pierced occasionally by a beam of light and blinking reds and greens. The slip-slap of water under the bow is a rhythmic staccato, popping counterpoint to the rumble and hum of the diesels down below.

At the end of his shift, Tom Flowers hands control of the vessel to the pilot, and he heads downstairs for a bite to eat. He tells the second mate he did a good job this evening. Then he heads to his quarters.

He calls home. He talks with Cathy and asks about Sam and Sarah. It's past their bedtime. He's been gone less than twelve hours, but he tells Cathy he misses her.

Later, while he's just about to surrender to sleep, he thinks about where he is. And maybe he'll cal-culate distance and time and try to figure out where he'll be in the morning. Wherever it is, he knows, it will be somewhere else.

Pushing a Quarter-Mile of Steel

Heading Upstream near Peoria

Not all of the narrows are this wide. Some of the bridges don't give a tow this much room. The whole rig is almost 1,200 feet long and 110 feet wide and weighs more than 22,000 tons. The boat's not going very fast, maybe three miles per hour, but it still takes a while to stop. One bridge gives only four feet of clearance on either side; it's like threading a needle.

Even with all the equipment in the pilothouse—the radar, sonar, charts, and reports—a pilot will tell anyone that his most valuable asset in a tight squeeze is the crew on the point. Connected by radio, they tell the pilot how close he is on the port and starboard.

"They're my eyes out there," Captain Tom Flowers says. "If you've got a good crew, you can put the tow anywhere."

He works the port engine, tamping down the throttle just a notch. And as his fingertips gently tap the "sticks" that control the rudders, a quarter-mile of barges responds.

Jason Ellis, Mike "Ranger" Riley, and Kevin Cluck after breaking the tow and reassembling and then passing through the La Grange Lock and Dam near Beardstown. River Mile Marker 81, August 2001.

Night Watch

A Break for the Deckhands

For now, the doin' is done. The tow's locked through, northbound on the Illinois above La Grange. The deckhands on the front watch finally catch a break.

The boat's got stops and drops in Peoria and La Salle, Lockport and Lemont. She's hauling salt, molasses, and chem—styrene and benzene and don't smoke out on the tow, boys.

The deckhands keep it all together—cranking it down, cinching it up, pumping out the holds. And they bust it down, too—locking through on a double-down, banging metal, cutting loose, checking off. And then they put it all back together.

Life is a whisper out there on the tow. You can trip and fall and drown in a minute. A line can slip and snap and take your leg clean off. You thank God for good boots and steady feet and a pair of White Mule gloves. Your muscles and ropes and steel cables are your stock-in-trade, and you get used to the grease in your jeans and the rust and oil on your skin.

But when the doin' is done, you can catch a break, have a smoke and a Coke, and maybe lie for a minute among the ropes. Your luxuries are counted that way. And you wait for the shift change. Six hours on, six hours off; thirty days on and not enough off; and the days on the boat, they never really end.

In a Tight Spot

The Pekin Wiggles

The spotlight scans the shore, feeling for buoys and landmarks in the dark. The pilot trains the beam on a familiar tree, makes a mental calculation, and nudges the starboard throttle.

Gauges and computer screens in the pilothouse are blinking and spitting out data—depth finders on the bow, channel markers, speed indicators—but nothing can replace the information the pilot reads with his eyes on the shore.

This is a particularly tight place to navigate when pushing this much floating steel. The channel narrows as it sweeps sharply to the right and then makes a wicked bend to the left. There's a dock sticking out into the river at this point, and the channel makes another hook to the right before it's over. It's rare that a tow doesn't go aground through here. This is the stretch of the Illinois Waterway known as the Pekin Wiggles.

It's possible to navigate this passage cleanly with a tow of fifteen barges, but the current has to be just right—not too swift, not too slow—and the water level can't be too low. And the pilot has to hit it at just the right angle, flanking at just the right speed and cutting the engines at just the right moment. And it helps if he holds his mouth a certain way and thinks positive thoughts . . .

Captain Tom Flowers knows he's not going to make it this time. The tow is barely moving as the

At sunset, looking south at the upstream end of the Pekin Wiggles with Powerton Power Plant smokestacks in the distance. River Mile Marker 151, January 2001.

lead right barge noses into the bank at the second turn. All engines halt. The spotlight sweeps the shore again, and Flowers goes to work.

With the deftness of a forklift surgeon, he swings the rudder sticks to the side and works the throttles that control the twin diesels below. Reverse all, easy does it . . . dead right. Okay now, power hard to port. . . . The towboat shudders as it cranks the barges off the bank and back into the channel.

The captain swings the floodlight along the shore. He's watching his rear now. All clear. And then he calls for more power from the starboard engine; forward now. He nudges the throttle. The bow of the barges slowly swings to port, and with a little more juice from the diesels, the entire tow clears the bend.

Engine Room Sanctum

The Chief Keeps It Running

You gotta keep it going. You're pushing fifteen barges full of cargo and you can't afford any downtime. That's your job if you're Jack Mandrelle, twenty-seven years a river man, chief engineer on the *Orleanian*.

"The name of the game is to keep it running," he says. "If you stop in the middle of a run, you're losing money."

The twin-12s—the big GM diesels that power this boat—can pull 52,000 horses, and when they're running full bore, the temperature will hit 120 degrees in the engine room and the sweat pours and it's all a roar and you can't hear yourself think.

And there's more.

Jack has a pair of Detroit diesels to worry about, too. Those V-8s send auxiliary power to the boat for lights and radios and wheelhouse outlets. And there are half a dozen supply fans and pumps to keep up and electric motors to maintain and countless filters and belts to change. Keep it running, Jack. You can't afford any downtime.

The engine room is loud, but he likes what he hears. It's hot, but he doesn't seem to mind. "It's a good boat," he shouts over the din.

He's grinning like a new daddy and says something else, but it's lost in the roar.

Engineer Jack Mandrelle in his engine room aboard the Orleanian. *November 2000.*

Keeper of the Galley

Aboard the Orleanian

After fighting currents or lugging chains or sweating over those diesels down in the engine room, a crewman can develop quite an appetite. And it's a lucky hand indeed who walks into the galley and finds Sondra Hooker putting food on the table. She's running the kitchen on the *Orleanian* on this particular trip, and her home cookin' is a favorite of deckhands all over this river and the upper Mississippi.

Crews for each run are assembled by the home office. Deckhands, pilots, engineers, and cooks are interchangeable parts. Sometimes a pilot and an engineer, or a first mate and a captain, might stay together for years on a particular boat, but bodies are frequently shuffled around. A deckhand often won't know what boat he's assigned to until he gets back from leave. It could be the fine clean boat or a stinky old tug with leaks and a grumpy cook.

But if a deckhand comes aboard and finds Sondra Hooker in the galley, he knows he's going to have a good run.

Towboat cook Sondra Hooker in the galley aboard the Orleanian. *The river and the shoreline can be seen through the screen door. November 2001.*

Where in the World

Morning, Somewhere on the River

Just waking up on the boat can be a little disorienting for the new guy. Time has passed, and he's been moving while he slept. He doesn't know where he is.

A person who works the boats long enough gets used to this feeling, and for the deckhand, it doesn't matter how far up or down the river he is. It only matters if it entails work. Are we near a lock? A narrow bridge? Any fleet work to be done?

For the most part, though, it doesn't matter where one is on the navigation chart, for one is always "here." In the existential moment. Here. On the boat.

The new guy sips a cup of coffee in the galley one morning, trying to adjust to the shifting landscape. He wonders where, geographically speaking, he is.

Mike "Ranger" Riley, the second mate, strolls into the galley and starts scrounging for a cup.

"Riley, have any idea where we are?"

"Well," Riley says, peering out the port window to get his bearings. "Right now, I'd say we're about 100 feet from the bank."

Near Seneca, looking west from the stern of a towboat. The sun sets downstream. River Mile Marker 251, November 2001.

The Boat Market

Recharging in Mid-run

The fleet of barges has been tied off against the bank, and the crew breaks it free from the towboat. Captain Tom Flowers watches their work from the wheelhouse and prepares to back the *Orleanian* into the channel again. It will be nice to be untethered from those barges, even if it's only for a quick trip across the river.

Being free from the burdens of work, even temporarily . . . there's nothing like it.

The *Orleanian* dropped its barges because it will have its tanks of potable water refilled at the Hennepin Boat Market, and it will have to dock for that. While its water supply is replenished, the boat will also have its galley restocked with groceries, and the crew can shop at the store and stretch their legs on solid ground.

Water, groceries, and shore leave—it's difficult to say which is most important, but one thing is certain: the *Orleanian* can get it all here.

The Hennepin Boat Market is about midway through what can be, for a deckhand, a monotonous run up the Illinois River. Operated by the Judd family for two generations, the market is a diverse operation with a warehouse downtown and a grocery on Front Street. At riverside, the business maintains a floating store on a tug called the *Robert J.*, which can bring supplies to towboats in midstream. Commercial haulers like the *Orleanian* call in

their orders—anything from groceries to a crewman's prescription order—and the market will have the goods waiting by the time the boat pulls in.

Water and groceries are the primary concerns of the engineer and the cook, but for most of the crew, the attraction is the *Robert J.*, a tiny craft with a cabin crammed full of goodies. It is floor-to-ceiling candy, chips, and chewing gum; T-shirts, caps, and comic books; personal hygiene products; men's magazines, videos, cigarettes, and novelties. A crewman can find just about anything he needs, or thinks he needs, right here.

Each person aboard the *Orleanian* has his own reason to look forward to Hennepin. They've all been here before, and each one has been anticipating this for days. For some, it will be another two weeks before they see family and home again. This is a place they can recharge their batteries, lighten the load, and loosen themselves from the burdens of worry and work.

Captain Flowers guides the towboat across the current, swinging it around to approach the dock from downstream. While he feels relieved to be refilling the water tanks and restocking the pantry, Flowers knows how important this stop is to the morale of the crew. After two, three, or in some cases four weeks without a break on shore, this stop is a precious thing. It can bring a smile to the greasy face of a deckhand. When they arrive at the dock, all hands are on deck.

Hennepin Boat Market employee Brad Vice aboard the Robert J. River Mile *Marker 207.5, July 2001.*

Two days earlier . . .

Ted Hartfelder is the cook on this run, and he pokes around the shelves of the galley, a checklist in hand. "Plenty of potatoes," he thinks, sticking his head into the bins of dry goods near the door of the galley. "Short on soups . . . we'll need some rice."

It's the cook's job to make sure there are enough groceries—from canned goods to fresh fruit, paper goods, and condiments. He's preparing his order for the Hennepin Boat Market.

"We'll need juice and breakfast cereal," he thinks. "I'll bet they'd like some brownies . . ."

He matches the goods on hand to his planned menu. Fish on Friday, steak on Saturday, chicken on Sunday—those are the givens. Cooks have a lot of leeway on what they serve a crew, but they dare not vary the dinner lineup on weekends. That menu is an unwritten tradition on the river, one that goes back years.

"Doesn't matter what boat you're on," he says. "Everyone on the river is eating fish on Friday and steak on Saturday. Guarantee it."

He's relatively new to the *Orleanian*, and the crew is still getting used to his cooking. He's getting used to them, too. He doesn't want to mess up on his first tour. He repeats the menu as if it's a mantra.

"Fish on Friday, steak on Saturday, chicken on Sunday . . ."

A day and a half earlier . . .

Roger Chilton, the first mate, steps onto the landing outside the wheelhouse and raises the binoculars. His smile broadens as he scans the beach.

"Let me see those," Greg Graves says. Chilton ignores him.

Three pleasure boats are nosed into the beach off the starboard bow, and half a dozen couples are partying in the sand. As the towboat passes, two of the women lift the tops of their bathing suits and flash the crew.

Graves whistles, Chilton smiles wider, and the pilot sounds the horn. Thank you, ma'am.

The Pekin sandbar is well-known to river men. On sunny summer days, young pleasure crafters fill the remote beach, inaccessible by land. Inhibitions, not to mention bikini tops, are dropped by the wayside. Flashing the towboat crews is considered safe fun by some of the young people who play on the river.

"They know we can't stop," Chilton says, not moving the binoculars from his eyes. "And they're trying to impress their boyfriends."

The first mate has a wife; Graves, a deckhand, has a girlfriend. It will be two weeks before they'll be home again.

The day before . . .

Captain Flowers radios ahead. He tells the market the *Orleanian* will be refilling its water tanks this stop.

The boat's engineer, Jack Mandrelle, monitors the boat's drinking water supply. He'll have one fewer thing to worry about when they leave Hennepin, but he has another concern, too, a personal one.

After twenty-seven years in the service of riverboats, Mandrelle is nursing a few old ailments, and today his supply of penicillin is running low. One doesn't want to run out of medication while on tour. He asks the boat market for a favor. With the market acting as the go-between, Mandrelle and a pharmacy conduct business. The prescription will be waiting for him with the ship's food order.

This is standard operating procedure for the market. A year earlier, Mandrelle's assistant, Tamer Gurmen, had developed a severe toothache in the middle of the run. The Judds made an appointment and shuttled him to a dentist in nearby Spring Valley. Then they made sure he rejoined the *Orleanian* upstream.

On this trip, Gurmen has no emergency, but he does have a special request from the market: five cans of Kodiak chewing tobacco, wintergreen, long-cut.

The *Orleanian* sidles up to the dock next to the *Robert J.*, and the deckhands quickly tie up. They form a bucket brigade to transfer the bulk food items to the boat: ten pounds of ground beef, twenty pounds of ground chuck, three fryers, one hen and a turkey . . .

They work quickly. The faster they get the spoils aboard, the sooner they can get into the candy store. Twelve pounds of pork sausage, three pounds of polish sausage, ten oranges, six pounds of bananas, five pounds of peaches, pears, and plums.

Hartfelder checks his list: two gallons of Crisco, three bottles of ketchup, eight cans of tuna, ten dozen eggs, onions, broccoli, cauliflower, lettuce, three heads of cabbage, green pepper, yellow squash.

The boxes are carted aboard and stacked in the galley. A clipboard and invoice in one hand, a pen in the other, Hartfelder tries to keep up. There's too much coming in at once . . . Parkay, swiss cheese, cottage cheese, sour cream, frozen catfish.

T-bones . . . where are the T-bones?

Mandrelle checks the water intake, and then he steps ashore. Stretching his legs, he glances back at the wheelhouse. Flowers is still up there. Someone hands Mandrelle his prescription. "Thank you."

Hartfelder has lost track of the order, but he's located the steaks. He'll sort the rest out later. Apple juice, orange juice, Cran-Apple . . . did we get the V-8? It'll turn up. The last box is aboard, and the deckhands head to the *Robert J.*

While the crew is in the floating store, Flowers starts preparing the boat to leave. In half an hour, the water tanks are filled and the food is stored. Hartfelder has located the V-8.

The deckhands are laughing as they carry their personal booty back aboard. There are fresh cigarettes and paperbacks, bubble gum and a pocket knife. Chilton has a video. He doesn't say what it's rated, but he promises to have a private viewing one night. Graves cradles an armful of magazines. There are a lot of pictures in there. Gurmen opens a can of Kodiak.

Flowers has been watching the crew board. In a few minutes, they will head across the river to pick up the barges and resume their run up the Illinois. It's good to know there's water in the hold and food in the galley. And it's good to see a smile on a crewman's face.

The Smell of the River

Chief Engineer, the Karla

One day in midsummer, the *Karla* is docked at the Hennepin Boat Market, and a stocky man with a hearty laugh walks the deck as if he owns the place. For all practical purposes, he does.

Bill Alexander is the *Karla*'s chief engineer, and while the captain is off the boat, Alexander is in charge. It is no big deal for him, no great added responsibility; he's run many a boat in his day.

"I was born on the Tennessee River, live on it now," he says, offering a quick tour of the *Karla*. He's built like he might have played fullback about thirty years ago, and he looks like the type who wouldn't duck a brawl if one came within a mile of him. A person would want to stay on this man's good side.

"I've worked the river all my life. I've been a pilot and deckhand, an engineer; heck, I've even been a cook on a boat."

The *Karla*, docked to take on water and supplies at the boat store, has a retractable wheelhouse that can drop hydraulically to the level of the deck, allowing the boat to sneak under low bridges.

Bill Alexander, chief engineer of the Karla, *near Hennepin, facing upstream. The Interstate 180 Bridge is visible in the background. River Mile Marker 207, July 2001.*

"These kinds of boat were built for the bridges in Chicago," he says. "You don't have to wait for them to open; you just scoot right under them."

The hydraulic lift works smoothly, gently raising and lowering the pilothouse. But a pilot has to stay on his toes, Alexander says. He doesn't want to forget to lower the pilothouse and wake up in the river. That's why there's an emergency drop.

"It'll take you from thirty feet in the air to zero in three seconds. Those thrill rides at the carnival ain't nothing." He grins mischievously, which might make one wonder whether he pushes that emergency button just for the heck of it every now and then.

But just when he seems like a fun-loving, rabble-rousing old bloke, he goes and gets soft, talking about the river like some rough-cut poet of the waterway.

"I smell the river when I get up in the morning, and I smell it when I get home," he says. "People don't realize it, but the river has a smell all its own. It doesn't smell like the ocean or the lake; it smells like the river. Nothing like it. And I don't like being away from it."

A Man and His Toys

Lock Operator at Work and Play

The enormity of this place can dwarf a man. Bolt heads are the size of fists, chains as fat as thighs. Heavy hinges pop and metal grinds. There are huge forces at work.

Randy Mayse walks along the lock wall at La Grange and watches the tow move in. He's been a lock operator with the U.S. Army Corps of Engineers for the past several years, and he's still impressed by the magnitude of this operation.

A rig like this—fifteen barges and the towboat—is almost 1,200 feet long and 110 feet wide. It's moving at a glacier's pace as the captain, sitting in the pilot-house three stories above the deck, inches more than 20,000 tons of steel and cargo into position.

As Mayse keeps an eye on the maneuver, his radio crackles to life. "This is La Grange; go ahead," he says. He's an affable man, and his friendly voice is appreciated on the river.

The captain of an unseen towboat downstream wants to know about the conditions at the lock and what might be coming his way. He'll be at La Grange in about an hour, about nightfall. Mayse tells him a big tow is locking through right now, but nothing else is in sight upriver. "You a single or a double?"

He's a single. That's good, Mayse thinks. Should be an easy pass. "Bring it in anytime you want."

If the barge industry had a benevolent uncle, one who showed up at Christmas with a sackful of smiles and plenty of toys, it would be Randy Mayse. He is quite literally the provider of toys to the industry. "I'm the only man in the world who makes toy barges," he confides with a touch of pride.

At La Grange, Mayse is surrounded by seemingly irresistible forces and immovable objects, but when he gets home he immerses himself in a tiny world of his own creation. With a set of molds, a batch of epoxy resin, and a bank of ordinary kitchen pressure cookers, he makes miniatures of the heavy equipment that dwarfs him all day.

The 1/600th-scale boats are detailed replicas of the real deal. At that scale, each barge is almost four inches long; a fifteen-barge tow, with boat, is about twenty-three inches long. But this is more than a hobby. It's developed into a small business, which he calls Cherokee Barge and Boat. When he's in full-production, he can pound out hundreds of boats a day.

"They're representations," he says, "short stubby copies."

The multicolored barges come in all types. There are chemical barges and open coal barges, covered container barges and barges that are only half-covered. They come with swept bows and boxed bows. And the towboats are different, too. Want one that represents a retractable wheelhouse? No problem.

He's made accessories that include full locks and dry docks. He makes tiny buoys and mooring cells. To add an element of realism, some of his mooring cells show dents where they were clobbered by runaway barges, and some have trees sprouting from them, a common sight along the waterway.

Mayse grew up along the Mississippi River, near the town of Louisiana, Missouri. After a stint in the

Deckhands guide a barge load through the La Grange Lock and Dam near Beardstown. River Mile Marker 80.2, April 2000.

service, he worked construction in Montana and other places, but he was constantly drawn back home to the Midwest.

"I've just always been fascinated with the river and this industry," he says. "The Illinois is such a neat river. You never know what's coming around the bend. I love everything about it: the boats, the people, the equipment."

It's the bigness of the river—and all that goes with it—that has always awed him. And the passion he has for his day job translates to his moonlighting work. It shows up in his toys.

"Each company on the river has its own logos and colors, and I customize the boats," he says. With a computer program, he designs a specific boat's markings, from the vessel's name on the pilothouse to the windows, lifesavers, and axes along the deck. He prints those full-color images onto an adhesive sheet, and they are then applied to the tiny boats.

Obviously, these are more than children's toys, although that's what they started out to be. But the industry discovered them and saw new applications for them. Mayse has sold thousands of units to barge companies. He says he's made 15,000 boats and 250,000 barges over the past four years.

The barges have been used as promotional material when a barge company is courting prospective clients. They are used in corporate presentations and as sales incentives. And they're used as training aids, to instruct pilots how to pass another boat, for instance, or to teach deckhands how to make up a tow.

They've even seen time in court, serving as props to help decide cases involving barge accidents and liability issues.

"They helped prove a case down in Tennessee once, and after the trial all the lawyers wanted a set of boats," Mayse says. "I sold twenty sets out of that case."

Despite the popularity of his boats within the industry, he knows there isn't a big market for them. It's a select clientele. As he says, "When you get twenty miles away from the river, you won't find too many people who know anything about the river."

It's doubtful they would know what a barge even looked like. No, he won't be making big bucks with his little enterprise, but that's all right with him.

"I haven't made a lot of money at it, but I sure have had fun. If it were really lucrative, I wouldn't need this job."

And then he laughs. Good old Uncle Randy. He'd miss it if he didn't go to work at the lock and stand next to those huge towboats and feel the power of steel, to look up and wave the captain through. The size of the place helps him put things into perspective, reminding him how small he really is.

And if that thought ever bothers him? Well, he can always go home and pull out the toys. There's a whole fleet of riverboats and barges awaiting his command.

Menagerie at Dresden

Keeper of the Lock and the Flock

Pauline Zitzke reaches into the black trash bag and pulls out a chunk of stale bread. She is trailed by half a dozen Canada geese.

She breaks off bite-size pieces and talks to the geese as she walks. "Here you go; come and get it." A few more geese gather in front of her. They know the routine.

These are her babies, along with the songbirds and the squirrels and chipmunks. She once helped raise a deer on these grounds. When she's feeding the animals, the weight of the world is lifted. There is peace here. She doesn't have to think about the boats, about the concrete and steel, about the cargo and those boys—for they are just boys, you know. Worrying comes easy on the river.

When she's finished feeding her menagerie, she goes back into the building, slaps the breadcrumbs from her palms, and gets to work. In a minute, she'll be at the controls again. She'll study the computer screen, banter with a pilot over the radio, and then press a button that dumps 10,000 gallons of water. In another minute, she will crook her finger at another switch, and two massive gray doors, each weighing more than 200 tons, will swing open to welcome a dozen fully loaded barges and another crew of river men.

U.S. Army Corps of Engineers lock operator Pauline Zitzke on the upstream side of the Dresden Island Lock and Dam. River Mile Marker 271.5, January 2001.

Zitzke is one of the lock tenders at Dresden Island Lock and Dam, the first woman ever to control a dam facility on the Illinois Waterway. She's been working with the Army Corps of Engineers since 1968 and at the locks since 1973. She's been at Dresden since 1992.

When she's on, she takes river readings every two hours and fills the pool and dumps it, sending boats on their way either upstream or down. It doesn't matter how big the boats are, either. "We locked through a guy in a kayak last summer."

The pilots and captains know her well. Some of the deckhands do, too. They have sort of a mutual aid agreement with her.

"They drop off big bags of old bread for the birds," she says. "And I keep them in touch with the outside world. I give them magazines and newspapers and sometimes vegetables from the garden. You wouldn't believe the buckets of beans I brought up here last year."

The crews sense something in Zitzke that goes beyond a friendly voice and warm smile. She genuinely cares about the crews, about the people who run up and down this waterway pushing impersonal product for invisible clients. She adds a touch of humanity to this run.

Her tone gets serious while telling the story of the young deckhand who got his leg caught between a taut rope and a kevel. It happened somewhere else on the river, during a lock-through just a week or two ago. It cost him his leg. "Sad, sad thing," she says.

Yes, worrying. It comes easy on the river.

Tenuous Banks and Roots

Standing Up Against Time

Rows of precarious trees line the banks in places, leaning out over the water as if pleading to passing boats.

The banks of this river used to be thick with oak and ash, but now mostly cottonwood and willow grow here, another face of the changing landscape.

And you wonder how much longer the river-soaked roots of these particular trees can hold up against the relentless weight of time. How many more years will their entreaties go unanswered?

New Look for Henry

The First Lock on the Illinois

In 1872, the first lock and dam on the Illinois River was completed in Henry. That dam did at least two things: it helped ensure a navigable channel downstream from La Salle, the terminus of the I&M Canal, and it pumped life into the little town on the western bank of the river.

Before the lock was built, the water was a slow, shallow current that suited the pace of the people here. But once the lock went in, the town took off and thrived until the coming of the modern Illinois Waterway in the 1930s.

The old dam was ripped out and the lock dismantled. Remnants of the lock remain today and serve as a draw for tourists and a reminder of prosperous times when the big boats docked and the whole town came down to see.

A long exposure with tree branches in motion during high water at dusk near the Pike County Fish and Wildlife Area. River Mile Marker 58, October 2001.

The disused nineteenth-century lock, which sits on the west side of the river, north of the Henry Bridge. A 360-degree image with the center looking east, the far left looking upstream, and the right looking downstream. River Mile Marker 196.2, November 2003.

Survivors in Peru

Down to the River

When Pat Shea came home from Vietnam, he cast around for pieces of his former life. He went down to the tavern that sat by the old railroad car.

Some of his buddies were still in town. A few had started families and had steady work. There were plenty of jobs in La Salle and Peru in those days. Shea went back to work at the family dry cleaning business. Everything was sort of familiar, but nothing was quite the same.

In the army, Shea was a crew chief on a helicopter. He saw action along the Cambodian border in 1968. And when he got home, it was hard to tell whether the town had changed or he had.

Either way, it didn't really matter. There had been too much water over the dam and under the bridge. Too much time had passed.

Meanwhile, work at the cleaners was slowly driving him crazy. It was a family business. He was the manager, the heir apparent. But after Vietnam, working inside all day was maddening, claustrophobic.

And then one day, he just walked.

During this time, there was one bright light in his life. Her name was Mary Mertel. She was a pretty brunette, and the two of them just clicked. She was easy to talk to. She seemed to understand what was going on inside him. They were married in 1973.

Mary's father, Joe, was one of the Mertel brothers who ran the gravel company down at the waterfront in Peru, and Joe knew his son-in-law was drifting.

*On Father's Day, Pat and Mary Shea at
the Mertel Gravel Company dock, Peru.
River Mile Marker 221.2, June 2002.*

He offered him a job and wouldn't take no for an answer because he knew Shea would get more than a paycheck out of it. Shea remembers the conversation.

"Come down here and get on the river," Joe told him. "You'll be busy the rest of your life."

And that's exactly what happened. He's been busy ever since. And Joe was right about something else, too: Shea got more than a promotion and a paycheck out of working on the water, more than you can put a price tag on. Mary had provided a purpose, and the river gave him focus. And it came at the time in his life when he needed it the most.

Mertel Gravel Company was founded by three brothers: Joe, Art, and Tony. It initially was a ready-mix business that pulled stone from a quarry on the bluffs above the river. But when the quarry played out, the company moved its operation to the riverfront to be closer to the material it now had to ship in. It was a matter of survival, but it ended up paying dividends.

The move to the river not only gave the company easy access to all types of materials but also presented other opportunities. It went from a ready-mix outfit to a company that could pour footings and build basements. It became a construction company, too. Then it diversified even more.

"One day they just decided they needed a boat, so they built their own," Shea says. "Right here on the side of the river. They waited for high water, and then they pushed it in."

That was the big step for the company. Now it could haul in its own material and haul material for others. By the time Shea joined on, the company had a second boat, this one built by a St. Louis company, and the river operation was starting to boom.

Shea initially worked as a helper on the new boat.

"I liked being a deckhand," he said. "You're young and strong then, and your knees aren't bad and your back isn't bad."

After a few years, the pilot got sick, and Shea was asked to take over in the wheelhouse. The old army training kicked in, and he threw himself into the breach. "You just end up doing it, that's all."

Shea just ended up doing a lot of things. Today he's in charge of the entire river operation for Mertel Gravel Company. It's a small company that has managed to survive because of its proximity to the river.

"We can compete against the larger ready-mix companies because we don't have to pay the transportation costs they do," Shea says. Earlier in the day, he had pushed two barges of sand to the plant. "That's the equivalent of 180 semi-truck loads if we had to have it trucked in."

Mertel now operates one boat, the *Triple M*, and Shea is the captain. The original boat was retired years ago. The company owns fifteen barges, and when it's not hauling its own stone and sand, it could be hauling coke or steel for other companies.

"We've loaded and unloaded just about everything," he says. "You name it."

The company has done some dredging, too. A bucket and crane fitted onto a flat barge does the heavy lifting, dragging up silt from the floor of the river and depositing it into a second barge alongside.

Mertel has dredged marinas and loading docks and the areas around intake pipes.

When the old Shippingsport Bridge was demolished in 2002, the *Triple M* was at the scene, and its crane helped lift the steel span out of the channel after the explosives had dropped it into the water.

If Mertel were strictly a ready-mix operation, it's questionable whether it would have survived. Its competition in that industry is a team of Goliaths, but Mertel has managed to survive because, like its captain, it came down to the river.

"We believe in it," Shea says. "If it weren't for the river, this place wouldn't be here."

Answering the Call

Slow Day at the River Club

Sometimes it gets busy around here. But on other days, like this one, life is lazy and all a person has to worry about is a friendly poke in the ribs.

Ken Williams has owned the River Club in Joliet for a couple of years. It's a little joint without a lot of frills, located just off U.S. Route 6, a little-used highway south of Interstate 80. There isn't a lot of drive-by business, but by being right there at Brandon Road Lock and Dam, it sees a few boaters, especially on hot summer days.

"We see a fair amount of traffic from the river," Williams says, taking a few moments to chat at the far end of the bar. "If the locks are open, the boaters'll go on through, but when they're waiting to

Tavern/restaurant owner Ken Williams on the rear patio of the River Club on Railroad Street in Joliet. The bar overlooks the Des Plaines River just upstream from the Brandon Road Lock and Dam. River Mile Marker 286.6, March 2002.

lock through, they might come in. Sometimes it's a long wait."

Williams says he had no intention of buying a restaurant and bar on the waterway. After he retired as a mechanic at Commonwealth Edison, he started checking out his options. "After ComEd, I just started looking for a joint. This was open, and here I am." He just wanted to run an establishment anywhere. But now that he's here, he loves it.

"It's a really neat waterway, especially at night."

But it's afternoon now, and the place is pretty empty. The jukebox, as if recalling last night, is playing B. B. King, "The Thrill Is Gone." The phone rings.

"Hey," Williams shouts down to a regular at the other end of the bar, "answer that, would ya?"

"I ain't answerin' the phone," she yells back, giving him a sly smile. He knew she'd say something like that. A poke in the ribs.

"You wanna work here or don'cha?" Back at you, babe.

"Hell, I wouldn't work for you . . ." She answers the phone. It's someone who wants to know if the River Club is open.

Williams is a big, robust man with a friendly manner. He is the kind of guy who enjoys a joke, even if it's played on him.

"The best thing about being along the river is you get a lot of screwballs coming in here," he says, loud enough to be heard at the other end of the bar. "Like that one over there."

"I ain't getting that phone any more for you," she returns the fire.

He returns to his tale. "It's a neat waterway," he says. "A little while back, this old-timer came down the river in one of those Viking boats. It had a big head on the end of it and everything. It was something to see. He was taking it down to Florida, I think, or someplace. He had all the time in the world."

The phone rings again. Williams and the woman at the end of the bar glare at each other. She doesn't wait for him to ask; she just shakes her head.

"I hate that phone," he says under his breath. "I'm going to tear that phone off the wall." And then, to the woman, "Hey, after you get through answerin' that phone, get two Lites and a Bud. One of those Lites is for you."

She was already off her stool, feigning disgust, and halfway to the phone.

"Yeah," he confides. "She's a good one."

Collision of Currents

A *Tributary Swirl*

When two forces of nature collide, one bends to the will of the other. This can be seen where rivers meet and at the mouths of streams. The melding of these currents is marked in swirls and eddies because there is violence at those confluences.

As the mist struggles to lift against the rain, a creek's current meets the river's. And before the lesser current submits, it whirls in futile protest.

Swirling patterns are created during a rainstorm as the shoreline drains into the Illinois River near Bluff City in Schuyler County. River Mile Marker 108.2, February 2001.

From Mud Creek

A Trail of Sediment

This is where it starts, far from the Illinois River on a ridge outside Reilly, Illinois.

Nothing more than a high spot on the flat Vermilion County landscape, the ridge is barely noticeable. Nevertheless, this little bump on the horizon is an important geological structure. It defines the boundary of the Illinois River watershed. Rain falling on one side of the ridge will flow into the Illinois; on the other side, rain will find another way to the sea, through the Wabash River to the Ohio.

This is rich farmland, some of the most fertile soil on the planet. There's no telling how much of it has been washed away over the years and lies on the bottom of the river today. Since the Clean Water Act of the 1970s, sedimentation has replaced industrial and human waste as the villain in the valley. And this is where it begins.

A field drainage pipe runs out of the ridge, and the water flowing from it into a ditch along County Road 3800N is clear. Washing over a sandy patch on the floor of the ditch, it looks almost drinkable. At the beginning of the watershed, it doesn't carry much baggage.

A century ago, Illinois was dotted by small farms bordered by woods and sectioned into fence-rowed fields. Over time, they gradually gave way to larger farms, vast, treeless expanses of row crops, the tilling fields of the industrial-size farmer. In the mad rush for bigger, higher-yield farms, soil conservation was a foreign concept.

But we've learned a few things since then. When surface water was allowed to run off freely, it gouged deep, ragged scars across the land, carrying with it tons of topsoil, chemicals, and agricultural productivity. Now, farm fields throughout the Midwest are plowed in contours and drained by tile and pipes. Surface water is routed into ditches that lead to creeks and rivers. Concrete structures placed in ditches in this very section of the watershed have proved that silt can be sifted out of the creeks and ditches before it reaches the river.

Green buffer zones keep the banks of creeks and rivers intact and stem the flow of runoff. Collection ponds hold that runoff and release surface water to the rivers in a more controlled fashion. These are effective weapons in the battle to retain soil, reduce nitrogen content in the river, and keep silt out of the waterway. We've learned this much.

Government programs, both state and federal, encourage farmers to set aside land for these purposes, and many farmers do so voluntarily. Yet some agriculture producers—pressured by falling commodity prices, greater international competition for their products, and rising transportation costs—still clear their fields of hedgerows looking for another half acre or two, and they till to the creek banks. Without the buffer and the hedgerow, surface water flows unimpeded to the river. And so does the soil. We haven't all learned.

The drainage ditch near Reilly skirts the county road for about a mile and then cuts northeast across a field. Along the way, more drainage tiles and pipes link up, and the ditch merges with other ditches and then a stream, and all along the way the water picks up silt and gets brown, and somewhere in southern Iroquois County it all dumps into a larger stream with an unfortunate name, Mud Creek.

Mud Creek eventually slops into the Iroquois River, a major tributary of the Kankakee, which merges with the Des Plaines to form the Illinois River. And that's where all of that suspended silt ends up.

Houses are few around here, and there is little reason to maintain the roads. They're pretty rough. An iron bridge crosses a ditch on the way to the creek, its bed of boards clunking and bumping under the tires of adventurous vehicles.

The road winds along a field and a line of trees to a neglected concrete bridge that marks the end of the trail. Flanked by a twisted old hickory tree and a discarded washing machine, the bridge crosses Mud Creek and the road peters out on the other side, fading into two sticky ruts in a field.

Beneath the bridge, the water runs swift and brown. It had rained the day before. The banks of the creek have collapsed into the water in places, just as the Iroquois River's banks have eroded, and all of it gets tossed into the chocolate churning soup that rushes downstream. The creek kicks up a flotsam stick, which twists and turns and slips out of sight into the swirl.

Mud Creek is only one example of the systemwide problem. There are literally hundreds of streams like this one, and it's impossible to know exactly how much silt ends up in the river. Estimates range from 6.7 million tons a year to 14 million tons. But whatever the amount, that silt chokes navigation channels, clogs private marinas, fills in lakes, smothers wetland vegetation, and destroys spawning beds for fish and other aquatic creatures. It also exacerbates flooding.

Commerce, recreation, and the environment pay the price. An example of this is found in Mason County at Matanzas Beach, which was once a vacation hot spot situated on Matanzas Lake. The lake is a long, narrow backwater, but sedimentation has made it difficult to navigate in low water. Like a lot of the off-channel lakes attached to the river, large submerged mudflats have developed. Now Matanzas Beach is a collection of homes owned by people who keep their boats on trailers, and they harbor hopes for higher water.

The task of maintaining an open navigational channel on the Illinois Waterway falls to the U.S. Army Corps of Engineers, which spends millions of dollars a year on its dredging operations, most of which are contracted to private dredging operations.

Two primary methods are used to extract sediment from the river floor. One employs a large rotating blade attached to pumps and pipes floated on pontoons. The pipes convey the silt either to a disposal site onshore or to a hopper barge. The other method uses a bucket and crane to scoop the sediment onto a

Near Kingston Mines, the Merlin McCoy, *operated by TNT Dredging, is pointed downstream. Shown (clockwise from right) is the bow with the cutterhead raised and draining, a close-up of the sixty-inch-diameter cutterhead, and a view from the top of the boat as the pontoon pipes carry the dredged material to a location on the eastern shore of the river. In the last photograph, the floating pipes were temporarily disassembled to allow an upstream tow to pass. River Mile Marker 144.5, September 2003.*

barge. Other systems have been tried and still others are on the drawing board, but these are the two most commonly found on the Illinois.

Bringing the sediment to the surface is only the first step in the process. Disposing of the sludge is a more formidable task. Some of it is contaminated by chemicals and metals, primarily found in sediment from the 1930s through the 1950s, when there were few regulations governing what could be dumped into the waterway. Disturbing that sediment could release those contaminants again, so these areas are generally left alone. And when it is removed, that dredge requires special—and costly—treatment.

But even the environmentally healthy sediment needs a dumping ground.

The Corps of Engineers has acquired the rights to hundreds of sites along the river to take dredge deposits. Many of them are only an acre or two, and capacity is limited. Often the dredge is simply spread at the site and seeded. In some cases, the public is allowed to take the clean dredge as free fill soil.

The Corps of Engineers has partnered with different government agencies and private interests to explore other uses for the sediment. It's been used on golf courses and in agricultural research projects at the University of Illinois. It can be used along highways and in other state construction projects. It can help reclaim strip mines and abandoned industrial sites.

In a program called Mud to Parks, barges of dredged sediment are pushed 160 miles upstream from Lower Peoria Lake to the old U.S. Steel South Works on the shore of Lake Michigan in Chicago. The steel mill closed in 1992, and nothing remained but 17 acres of slag and the buildings' concrete foundations. The hope is that once it is covered with several feet of sediment from Lower Peoria Lake, a lush lakeshore park will emerge.

Another proposal calls for dredged sediment to be used to rebuild levees in New Orleans and other hurricane-prone areas.

There is a good reason why these projects are focused on the two Peoria lakes.

During a 1999 joint U.S.–Chinese workshop on sediment transport and sediment-induced disasters, David Ta Wei Soong of the Illinois Water Survey described the waterway at Peoria as "a narrow, deep navigation channel between flat submerged plains." The lakes—the Upper and Lower Peoria lakes—are barely two feet deep. In 1903, the lakes were estimated to have a volume of 120,000 acre-feet. In 1985, they had less than 40,000 acre-feet.

The lakes' depth makes them suitable for another sediment-fighting project, one that results in the creation of islands made of dredged material. The island-building project employs a large mesh cylindrical tube pumped full of sediment. At six to eight feet in diameter, the silt sausage rests on the bottom of the lake and rises well above the surface, and once the tube is linked end-to-end to form the outline of the island, more sediment is pumped into the center to build it up.

The price tag for building islands and creating parks is high and requires the commitment of a number of government agencies. But no matter how extensive the project, each has the feel of being a local solution to a much wider problem. The flow of sediment into the waterway continues.

Agriculture is not the only culprit. Housing developments on the ridges and along tributaries have replaced ground cover with rooftops and asphalt drives. Rainfall isn't absorbed into the ground in these developments. Instead, it is funneled very efficiently into ditches and routed downhill toward the river. This swift water contains a lot of suspended silt, and when it meets the slower Illinois River current, that silt drops to the bottom.

Coming Full Circle

Bridge Views

Spring Valley and Seneca are about thirty miles apart, but drawing them closer—although not in the geographical sense—are the bridges at both locations.

Erected about the same time, during the bridge-building boom of the 1930s, these structures bear striking similarities, especially when viewed from below through the lens of a 360-degree camera.

The highway bridges over the Illinois River from the north shores at Spring Valley (above) and Seneca (below), photographed in 360 degrees. The leftside half of each bridge is running to the north while the ends on the right extend over the river to the south. River Mile Marker 218.4 and River Mile Marker 252.7, respectively, July 2004.

Looking west into Chicago from within the Chicago Harbor Lock at the mouth of the Chicago River. Functionally, the lock at Lake Michigan is a water level control unit. The U.S. Army Corps of Engineers cites the facility as one of the busiest in the nation. River Mile Marker 326.7, August 2000.

CHAPTER THREE

Chicago's Reversal

Far beneath the streets of Chicago—350 feet below the Magnificent Mile, the Loop, and O'Hare, below the storm grates and gutters, down there in the twisting, dark abyss—a subterranean river flows.

This isn't a river you will find on any geographic map, but it could be the most important waterway in the city. It just might be the most important tributary of the Illinois Waterway. This is the Deep Tunnel: more than 100 miles of manmade, concrete-encased tunnels big enough in places to drive two buses through side by side.

This is Chicago's supplemental sewer, a safety valve, a long, tubular holding tank that relieves the city's antiquated storm sewer system. If not for the tunnel, a heavy rain would wash storm water and untreated sewage into the rivers. If it does its job, then the city, Lake Michigan, and the Illinois Waterway stay clean.

There are actually three separate tunnels, as big as thirty feet in diameter. They are part of the city's Tunnel and Reservoir Plan—or TARP—an ongoing wastewater and flood-control project that began in the 1970s and that eventually will include three above-ground reservoirs capable of holding billions of gallons of runoff. Managed by the

Metropolitan Water Reclamation District of Greater Chicago, TARP is the latest attempt to solve a problem that has plagued the city since it was founded in 1833: its own waste.

In the early nineteenth century, few towns gave much thought to sewage. If there were a ditch or a running river nearby, the offending effluvium would be routed in that direction, and the problem would simply float away. That's the way it was virtually all across the fledgling nation.

In Chicago, the river that meandered through town was a slow-moving stream, and it proved to be a fertile environment for incubating disease, especially when the population spiked. The construction of the Illinois & Michigan Canal in the 1830s and 1840s attracted thousands of people, and the little frontier town became a city virtually overnight. It grew faster than its infrastructure.

In the 1850s, the city constructed its first sewers, wooden and later brick structures that were designed to funnel rainwater runoff away from the roads and into the river. As indoor plumbing became common, human waste was directed into those same sewers. Although the sewers were improved and replaced, the

Between the McDonough Street and CSX Transportation Railroad bridges along the Des Plaines River in Joliet. This photograph was made from the concrete retaining wall running parallel to South Water Street, which is far below river level. River Mile Marker 287.5, June 2003.

concept of joint storm and waste sewers endured. The city was built on that foundation.

There was more than human waste being emptied into the river. A thriving stockyard industry—opened in the 1860s—started dumping its refuse, its carcasses and rot, into a fork of the Chicago River's South Branch, so fouling that stream that it became known as Bubbly Creek. All manner of industrial refuse and human and animal waste went directly into the river.

The final ingredient in this recipe for disaster was the fact that deep wells couldn't meet the demand of the growing population, so drinking water was drawn from the lake, which is where the river emptied.

As early as the 1840s, the situation had become deadly. A cholera outbreak killed thousands in 1854.

Thousands more died from cholera, typhoid fever, and diphtheria in the 1860s, in 1871, and in 1885. Almost 4,500 people died from typhoid alone between 1890 and 1892. Clearly, something had to be done.

LOOKING FOR A SOLUTION

The city moved its water intake cribs farther into the lake, away from the mouth of the Chicago River. And then it moved them again. But when it rained hard, the discharge from the river still pushed the pollutants beyond the cribs.

In the late nineteenth century, the city's engineers concocted a scheme to reverse the flow of the

Chicago River. Instead of having the river carry the city's sewage into the lake, the river would carry the filthy effluvium the other way, into the Illinois River, where it would disperse, dissolve, and be forgotten. The earliest attempts at this, made before some of the more deadly outbreaks of disease, involved the I&M Canal.

For economic reasons, the I&M was not dug as deeply as originally intended, and the elevation of its eastern terminus was higher than the surface of Lake Michigan. In order to get water into the canal, it had to be pumped up from the Chicago River, and boats traveling up the South Branch of the Chicago River to the canal had to be raised by the lock at Bridgeport. The pumps and the canal, therefore,

were moving polluted water out of the city, but not quickly enough.

So in 1871, the so-called Deep Cut was made. The canal was deepened, the lock at Bridgeport was removed, and the Chicago River was, for the first time, reversed. The water cleared at the lakefront, the stench was reduced, and the lake's water was safe to drink again.

It was an engineering marvel, a technological achievement cheered by everyone except those downstream. And it worked well. Until it rained.

Because the canal was simply too small to handle the flow, the Chicago River backed up whenever it stormed and again flowed toward the lake, taking its contaminants back to the intake cribs. The severity of

Looking northeast to the Chicago skyline from north end of the Damen Avenue Bridge, which crosses the Chicago Sanitary and Ship Canal, seen on the right. Chicago Sanitary and Ship Canal Mile Marker 321.1, April 2001.

the problem, coupled with a declining commercial value of the I&M Canal, led to the decision to create a new canal strictly for sanitation purposes. The Chicago Sanitary Canal was built during the 1890s and completed in 1900. It was almost three times as wide as the I&M and four and a half times as deep. And it was twenty-eight miles long, running from Chicago through Lemont to Lockport. Its purpose was singular: to reverse the flow of the Chicago River once and for all in order to flush the city's sewage downstream.

Although it would seem to be an obvious application of the new canal, commercial navigation was an afterthought, a fact that underscores the seriousness of the sewage problem. It would be six more years before the canal would be modified to handle commercial boats, and the word "Ship" was added to its name to reflect its new role.

The success of the Chicago Sanitary and Ship Canal led to the construction of a second canal, which forged another link to Lake Michigan, this one through the Calumet River. It connected to the Chicago Sanitary and Ship Canal near Lemont at a place known as the Sag, or Sag Bridge. The Calumet–Sag Channel, built between 1911 and 1922, provided a link between Chicago's new Calumet harbor and the Sanitary and Ship Canal, one that bypassed downtown Chicago. It too reversed the flow of a river, and it too moved sewage.

Both the Cal–Sag and the Sanitary and Ship Canal were integral steps in the eventual construction of the Illinois Waterway in the 1930s, a waterway that evolved, in at least one regard, from a sanitary ditch.

A QUALIFIED SUCCESS

The reversal of the Chicago River was hailed as an engineering success, one measured in lives—or rather, in the absence of deaths from waterborne disease—but the damage downstream was not included in that calculation.

The flow of city waste, both human and industrial, had an immediate effect on the river system downstream. Fish kills were common, and the stench emitted from the waterway was overpowering well past Joliet. The Illinois River had become an open sewer, Chicago's dirty ditch.

Wastewater treatment was unheard of at the time the rivers were reversed, and it was a novel idea in the 1930s, when the first sewage plant was built in Chicago. The basic operating principle has changed little through the years. Wastewater is routed through a series of filters and settling tanks. The water is aerated, fed microbes, and given time for natural biodegeneration to occur. No chemicals are added in this process. By the time it is emptied into the waterway, it is, in theory, clean. You might not want to drink it, but when the process works as it should, no fish would die from swimming in it.

The wastewater plants of that era had an immediate impact. But Chicago's integrated storm and sewage system worked against the process. Whenever it rained heavily, the treatment plants were overpowered and ineffective. Although the Chicago Sanitary and Ship Canal was big enough, for the most part, to keep the Chicago River flowing away from the lake, heavy rains still outstripped the treatment plants' capability. Several times every year, sewage had to be routed directly into the river.

The city lived with this situation for decades, but it was obvious that another solution was needed.

LIGHT AT THE END OF THE TUNNEL

It wasn't until the early 1970s when the city—prodded by the federal Clean Water Act and faced with the alternative of replacing its entire sewer system with separate storm and wastewater sewer lines—began construction of the Deep Tunnel.

The tunnel, which is actually a series of subterranean concrete-lined sewer pipes, roughly shadows the routes of the rivers that drain the surface: the Chicago, the Des Plaines, and the Calumet. Vertical shafts from the existing sewage system link to the tunnels, as deep as 350 feet below the surface. The tunnels lead to pump stations and wastewater treatment plants.

TARP is an ongoing project, and once the three above-ground reservoirs are on line, they will be able to hold 16 billion gallons of water, acting as holding tanks to relieve pressure on the treatment plants, giving the plants time to catch up to the demand. Once operational, the entire system should ensure that no

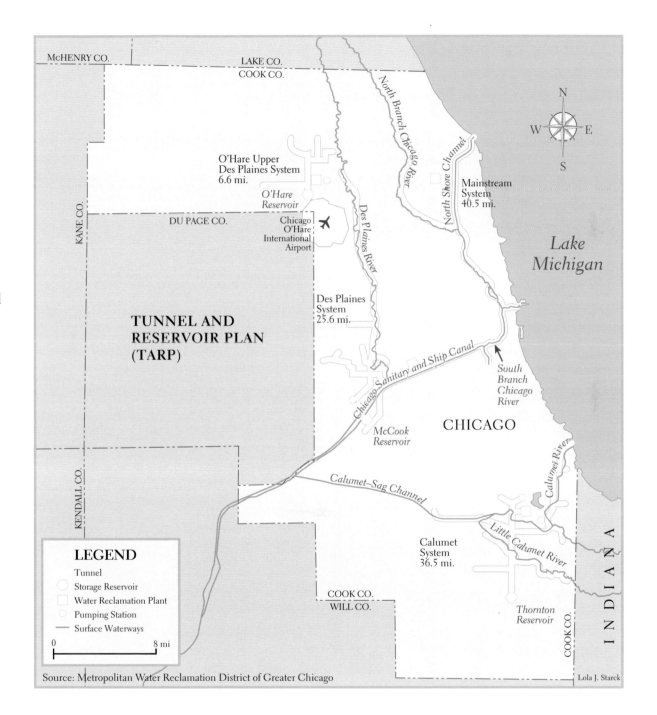

McHENRY CO.
LAKE CO.
COOK CO.

KANE CO.

DU PAGE CO.

O'Hare Upper
Des Plaines System
6.6 mi.

O'Hare
Reservoir

Chicago
O'Hare
International
Airport

North Branch Chicago River

North Shore Channel

Mainstream
System
40.5 mi.

Des Plaines River

Lake Michigan

Des Plaines
System
25.6 mi.

**TUNNEL AND
RESERVOIR PLAN
(TARP)**

Chicago Sanitary and Ship Canal

South
Branch
Chicago
River

McCook
Reservoir

CHICAGO

KENDALL CO.

Calumet–Sag Channel

Calumet River

Little Calumet River

Calumet
System
36.5 mi.

LEGEND
◯ Tunnel
◯ Storage Reservoir
☐ Water Reclamation Plant
◯ Pumping Station
— Surface Waterways

0   8 mi

COOK CO.
WILL CO.

Thornton
Reservoir

COOK CO.

INDIANA

Source: Metropolitan Water Reclamation District of Greater Chicago

Lola J. Starck

untreated sewage enters the waterway or backs up into the basements of the city's homes.

The effects were noticeable from the day the first tunnel was opened, and water quality has improved steadily ever since. The Chicago River, once a nuisance and health hazard, now has become an asset and a resource.

The city of Chicago has in recent years begun a concerted effort to revitalize its riverfront, rebuilding Wacker Drive and constructing walkways and parks along the water. Excursion boats regularly ply the waters, transporting camera-toting tourists down the river that once stank with stockyard and human waste. Water taxis operate and personal watercraft can be rented in Chicago. Real estate values have soared.

On the Cal–Sag, anglers can be seen crowding the banks at the mouth of the river. Bass tournaments are held on the Calumet, and sports fishermen take their boats all the way down the canal from the lake. Sometimes they launch their boats at ramps downstream. In Alsip, personal watercraft enthusiasts regularly drop into the canal. They can be seen zipping down the open, straight canal at forty miles per hour, water spewing into the air behind them.

The Cal–Sag's dark water still has the smell of a slow-moving stream, one that sees a lot of diesel and steel and treatment plant emissions, but it manages to support aquatic life. Playing a vital role in the effort to clean the Cal–Sag are the Sidestream Elevated Pool Aeration—or SEPA—stations. These manmade urban structures perform the function of natural waterfalls, adding oxygen to the water. There are five stations along the Cal–Sag with a combined capacity of moving 1.3 billion gallons a day.

Pumps draw water from the canal into a pool above the bank. In a series of steps, that water gradually falls back to the canal. The areas around these SEPA stations are groomed as parks, and they attract waterfowl, dog-walkers, and picnickers.

The areas along the Cal–Sag Channel and the Chicago Sanitary and Ship Canal somehow don't fit the profile sketched by veteran river men. They tell stories from the not-too-distant past of seeing rats scurrying across debris that floated in the Cal–Sag. Bank to bank, the rodents wouldn't wet a foot.

Those river men still refer to the Illinois River as a sewage ditch, but they call it that out of habit, not for what it is today. Old reputations are difficult to shake, no matter how many reversals are made.

At the Edge of Chicago

Where It Begins and Ends

Out here on the edge of Chicago, where the waters run together, one long finger of the city pokes into the cold Great Lake. It both points out to sea and beckons ships to land.

That is where, at the dawn of a different century, a regiment of teenage soldiers bivouacked before shipping off to battle. A thousand hopes were launched from here, and they sailed to a war that sank a thousand dreams.

A chained cluster of wooden
mooring timbers located
in the harbor between the
Chicago Harbor Lock and
Lake Shore Drive. Chicago
River Mile Marker 326.5,
April 2001.

Assistant Chief Operating Engineer James McCague and Chief Operating Engineer Henry Marks in the subterranean pump house of the Calumet Water Reclamation Plant. River Mile Marker 327.5, April 2001.

This spot was a destination, too, where merchants and immigrants disembarked and built Sandburg's City of the Big Shoulders. Lake steamers and sailing craft docked here, loaded with the building blocks of a nation. Or they pushed up the Chicago River and into the heart of the city, and in later years the cargo reached deep into Illinois and on to the Great Plains and the West.

This is a place of contrasts. This is the spot that fed a city and served a world. This is the place that launched fortunes—and empty promises, too. Many died here, but for some this is where their ship finally came in. And this is where the Chicago River once ended, and now it begins.

Pump House Gang

Calumet Pumping Station

Below the city of Chicago, the Deep Tunnel flows for more than 100 miles until its interconnecting passages reach the big yellow pump houses at the end of the line.

That's where guys like operating engineer Henry Marks and his assistant, James McCague, take over. They supervise the pump house of the Calumet district. Here, sewage is pumped at a controlled rate to the wastewater treatment plant above ground.

"These pumps can move about 200,000 gallons a minute," Marks says. That's a lot of waste, all of it under pressure and flowing past their heads every sec-

ond. One shudders to think of what would happen if something burst.

There's a claustrophobic, submarine-like feel to the place. Indeed, the pump room is sealed off by a hatch door with a porthole window. McCague puts it into perspective and states the obvious.

"If you look through the window and see fish," he says, "then don't open the door."

Confluence of Canals

A Sanitary Situation

The scene can be serene, almost idyllic, with its stone beacon and manicured lawn, geese lolling about.

But the waterfalls that flow both left and right are manmade, as are the waterways on either side. And none of them would be here if not for the need to move human waste. This is where the Cal–Sag Channel and the Chicago Sanitary and Ship Canal meet, and it's not always been this pleasant. Before it became apparent that dumping sewage straight into the river might be a bad thing, these two canals were open sewers, flowing cesspools of human and industrial waste. Nothing lived in these waters.

But today, Chicago's wastewater treatment plants are keeping up with the flow, and five SEPA plants along the Cal–Sag are raising dissolved oxygen levels enough so that aquatic life is rebounding.

This is the last of the SEPA facilities, all of which function in the same way, pumping water from

The entrance to the TARP pump house at the Calumet Water Reclamation Plant. Near Calumet River Mile Marker 327.5, April 2001.

the channel into an elevated pool and returning it through a series of small waterfalls. It's nature's way of suspending oxygen in water, which is vital to the survival of aquatic plants and fish—nature's way but manmade.

A Bobber on Bubbly Creek

Toxic Reputation

Visitors to this location have to pick their way past the broken glass and plastic trash, past the busted bricks and spent bait containers, down to the water.

The bank is littered with broken bottles and crushed cans, paper and pieces of sopping cardboard.

Those old shipping boxes still bear the logos of manufacturers and the U.S. Postal Service. Bad publicity, it is.

There is a stink about this place, but it used to be worse, a lot worse.

This is Bubbly Creek, the notorious fork of the South Branch of the Chicago River that once carried away the fetid waste of the infamous Chicago Stockyards. For years after the yards were opened in the 1860s, the gutted carcasses of cattle, the entrails of an industry gone mad, were dumped into this tributary. This water became so polluted, it literally bubbled. And the name stuck.

At the water's edge today, a lone fisherman squats amid the litter of the twenty-first century and keeps

The SEPA plant at the confluence of the Calumet–Sag Channel (left) *and the Chicago Sanitary and Ship Canal, upstream from Lemont. River Mile Marker 303.5, June 2002.*

At the intersection of Eleanor and Fuller streets, a rough platform sits in a weathered tree on the eastern bank of the once notorious Bubbly Creek. Visible across the water, toward Ashland Avenue, is the newly reconstructed western bank. Chicago River Mile Marker 321.8, March 2004.

an eye on the bobber about twenty feet from shore. He fields the question thrown at fishermen all over the nation.

"No, nothing yet."

Lester Poole, a neat, lean man of about fifty years, fishes these waters two, maybe three times a year, he says. There are all kinds of fish here: "Carp, bass, bluegill . . . just about everything." There could be a perch lurking under that plastic chair half submerged ten feet off the bank.

But Poole, who lives near 87th Street and Bishop—just down Ashland Avenue—doesn't much care what

he catches. He's fishing to relax. He doesn't eat what he catches here, and he says he doesn't know anyone who does.

"If I did know anyone who ate these fish," he says, "I wouldn't know them for long."

Although the Chicago River has been cleaned remarkably, there are still areas where the city's ugly past is too close to its present. This is one of those areas.

The city, lately, is giving the area some needed attention. A park is in place not far from the river's edge, offering a nod to the location's historical significance. This is where the I&M Canal began so many years ago and where the Chicago Sanitary and Ship Canal begins today.

This part of Bubbly Creek is a turning basin on the canal. It's wide and deep enough for boats to maneuver around in, but the creek itself is short. It doesn't get much circulation.

The stockyards have been closed for years, and the flow of organic waste into the stream was stemmed long ago. Still, the water can't be good here, and one wonders how fish—even the inedible ones—manage to survive.

In a few minutes, another fisherman slides down the bank and sets up about fifteen feet away. It's Ray Sullivan. He lives over on Harlem, and he's a veteran of these waters.

"Yeah, I eat them sometimes," he admits, slapping a ball of corn onto a hook.

"You're a mighty brave man," Poole says.

"You just soak 'em in salt water and deep fry 'em," Sullivan says. "Hot grease'll kill anything."

The cautionary words Poole spoke earlier bubble ominously to the surface.

The Clean Sweep

Picking Up Along the River

The rain didn't discourage everyone. Dora Dawson was out in the weather leading the charge for the annual cleanup. She was the captain for this section of the river.

Every year, hundreds of people in rain gear and gloves clamber up and down the riverbank of the lower Illinois, picking up the debris of the thoughtless. Chicago owns no monopoly on river litter. It can happen anywhere.

"You wouldn't believe the things they've found," she says. "Old refrigerators, tires, lawn chairs, you name it."

Dawson is the president of the Meredosia Historical Society and River Museum and a board member of the Friends of the Illinois River, a grassroots nonprofit organization that sponsored the 120-mile long River Sweep.

A broad cross-section of people gets down and dirty once a year to clean up the riverbank, including Boy Scouts, a group of girls from a local sports team, and adults from all walks of life. There are boaters, sportsmen, common laborers, retirees, teachers, and politicians. There's even a judge.

The Honorable Judge David Bone, circuit judge out of Jacksonville, took time out of his court schedule to participate in the cleanup. He's a regular around here, and he's Dawson's star witness, her chief volunteer.

"I've always been a picker-upper," he says, explaining why he came out in the rain to help hoist rusted barrels and trash out of the water. "When I was a lieutenant in the Marine Corps, I'd try to set a good example by picking up cigarette butts."

It must have been an odd sight in Vietnam, 1965, to see a young officer just back from recon patrol with

Circuit Judge David Bone and Dora Dawson during a downpour under the east end of the Meredosia Bridge on the day of the annual Illinois River Sweep. River Mile Marker 71.3, September 2001.

his unit, bending down to pick up a grunt's butt. But Bone didn't care. He did it anyway, he says.

The weather chased off a few volunteers, but there were enough people to collect a nice pile of trash. Dawson proclaimed it a victory. Still, it's hard to top that one year when they found three TVs, a dishwasher, and a clothes dryer with a load of clothes still in it.

Shifting Perspective

On Patrol with the Chicago Police Marine Unit

From the river, everything takes on a different hue.

"See that up there?" Ted Parker points to the top of the Chicago Board of Trade Building. Atop the forty-four-story art deco skyscraper, against the backdrop of an azure sky, is a statue of Ceres, the Roman goddess of agriculture. That it exists is not news, but from this vantage point it seems new and now strangely significant.

"From the street you don't even notice it," Parker says.

Sergeant Ted Parker is a thirty-year veteran of the Chicago Police Department. He spent ten years on the street before joining the marine unit, and he credits the Chicago River with changing his perspective on the city—and on life, too.

Being a police officer and going on patrol in a boat might sound like easy duty, and it's not surprising that there's a waiting list to transfer to the unit. But marine duty is no walk in the park.

As he relates this, Parker is piloting the thirty-two-foot cruiser from the unit's headquarters at Monroe Harbor around the breaker toward the intake lock that opens to the Chicago River. The lake is relatively calm today, but there are whitecaps, and the waves hammer the hull as the boat cuts toward the lock.

Marine cops are often on Lake Michigan, sometimes conducting rescue operations or working with the Coast Guard. And when the wind is up and the weather is cold, it's as intense as a running gun battle in the streets. Parker's been in tough spots both on the street and in the boat, and he'll admit, "I've never been more scared than out here."

Officers in the marine unit are trained for all types of situations. They could be called into action if a plane goes down in the lake or if a motorist drives his car off a bridge and into the river. They are all certified divers and go under looking for weapons or searching for bodies.

The mission today has an international flavor, and there are national security overtones. Mexican president Vicente Fox is in town, and his motorcade will leave downtown and cross the river on its way to the airport. Parker and his partner for the day, first-year man Dave Bryja, are sent to the Congress Parkway Bridge.

Their orders are to look for suspicious activity and halt all boat traffic under the bridge at the designated time. On the way there, they are to conduct a daily patrol. So while they are protecting one of the leaders of the free world, they are also to make sure no one violates the no-wake rule.

The water is calm once the boat leaves the lake and locks through. In the shadows of bridges, the Chicago River is dark green but for the most part clean of debris. Ducks gather near the pilings. A gull occasionally sweeps the area. Pigeons roost, of course; there are pigeons everywhere.

A solo rower sculls past the patrol boat under the Columbus Drive Bridge. Such a sight is not that unusual these days. The water is cleaner than it's been in more than 150 years.

"When the weather gets nice, everyone wants to be out on the water," Parker says.

Parker keeps his own boat, a twenty-nine-foot cabin cruiser called *Sudden Impact*, at a marina off 59th Street. He takes his grandkids joyriding as often as he can. Sometimes he fishes. And a lot of times while he's on the job, he can't help thinking about his own boat. Retirement is in his mind. Thirty years is a long

Ted Parker (left) and Kevin Williams of the Chicago Police Marine Unit, *tied off beneath the raised Union Pacific Railroad Bridge north of Wolf Point, where the branches of the Chicago River converge. Toward the south, behind the officers, is the Lake Street Bridge, which carries both automotive traffic and commuter trains of the Chicago Transit Authority's Blue Line. North Branch of the Chicago River Mile Marker 325.7, August 2001.*

time. A person needs his dreams, and Parker dreams a lot about his boat.

The sun dazzles off the white face of the Wrigley Building and bounces into the crevices formed by mirrored and monolithic glass buildings. It casts shadows deep and dark behind the friezes and turrets and illuminates architectural nuances twenty stories up. This is an old city that has managed to seamlessly fold hard, new structures into its fabric.

From the river, one can see how it all fits together. And the details are more noticeable — the iron work under the bridges, the walkways and rails down by the water.

An excursion boat named the *Star of Chicago* and a large pleasure craft slowly motor past. A water taxi is preparing to launch. The pilot waves as the patrol boat passes.

Under the Lake Street Bridge, the patrol boat heads south. The light changes dramatically, and the city shifts.

"Every time you come down here, you see something different. It changes, depending on the time of day," Parker says. "You should see it at night. It's stunning."

The patrol boat reaches the Congress Parkway early. Parker radios in and gets clearance to patrol farther downriver. He travels under Roosevelt Road, past industry, and there's Chinatown off the port bow. Just at the point where the river curves westward, a pleasure boat, a little runabout, comes skipping around the bend. It's going too fast and leaving a wake. Parker hits the police lights. Pull over, buddy.

The stop ends in a warning to the pilot, who happens to be a retired police officer himself. Cracking down hard on boaters is not part of the mission of the Chicago Police Marine Unit, Parker says.

"Most of the people who use the river love it," Parker says. "They're very cooperative. We're here to be of service, and in a situation like this, we'll remind them that each boat is responsible for its own wake, and wakes can cause a lot of damage in a narrow channel like this."

With the warning ticket issued, Parker takes the patrol boat back to Congress Parkway. Bryja ties up to a piling. There isn't much traffic on the water here, so closing the river won't be that difficult. The *Kiowa*, a small tow operating out of the Port of Chicago, pushes a Material Services barge full of gravel. Parker lets it pass and then closes the river.

Now they wait. The hard part.

Waiting can lull a person to carelessness. Waiting can give a person too much time to think. Parker thinks about *Sudden Impact* and his grandkids. And he doesn't want to, but he thinks about his old job, his time on the street. It was all a long time ago.

The attitude a police officer meets on the river, Parker relates, is vastly different from the attitude he confronts on the street, on the beat, up there in the jungle where you can get killed in a flash or watch your gut-shot partner die before an ambulance arrives. Those kinds of images can eat a man up from the inside. Parker knows all about it.

"You get surrounded by the negativity up there," he says.

Conrail bridges crossing southeast over the Calumet River with the Interstate 90 Chicago Skyway in the distance, near 95th Street in South Chicago. Calumet–Sag Channel Mile Marker 332, March 2000.

A beat cop's routine can turn him hard and mean. The constant grind of the beat can wear on him, and he doesn't even know it's happening. He picks up the radio and checks on Vicente Fox. Still waiting.

"A job like that can get you to build walls inside," he says. "Nothing gets out. It's not healthy, not for you or anyone around you."

Parker tells the story in second-person narrative, as if it's happened to someone else. But it's his story. He checks in on the radio again. Nothing.

The river has been a blessing. Its curative powers gave Parker a fresh perspective on life, this city, and the people in it. Once again he can think in terms of service to others, which was his motivation to become a cop in the first place. It was a motivation that ten years on the street almost beat out of him.

The radio cackles; it's HQ. Vicente Fox's motorcade has taken a different route. He's at the airport already; all clear at the Congress Parkway. Time to head back.

As he turns the patrol boat up the main channel toward the lake, Parker spots a mallard hen with four chicks in tow. They are trying to paddle across the river just up ahead. Parker powers all the way down, and the ducks pass.

Chicago Nexus

Bridges on the Calumet

From the first bridges that cross the Calumet, one can see where the river begins. This is the industrial leg of the Illinois Waterway. The Calumet is more Chicago than the Chicago. Both feed the Illinois with waters from the Great Lakes, but the Calumet pushes more freight. In the city that works, this is the one that does the heavy lifting.

Barges come this way, hauling salt and rock and scrap. Trains cross this way. These are the yards. This is where the tracks converge, where new steel meets old. This is where chain link fences and pilings and sheet metal with rusted bolts hold back the banks of the river.

This is where the laborers punch in, round the clock, and it never stops. This is the city of Chicago. And trains rumble and cross and merge in flat cinder lots, where boxcar numbers are scanned like computer code and counted by digital eyes.

This is where the freight ends, where it switches hands, where it lands on the docks to wait for the boats and the next leg of the journey. Downstream, on its way, miles to go, work to do.

One Man's Garbage Fish

Carp Fishing on the Chicago

It's a warm afternoon, and Paul Pezalla has parked himself in a familiar spot on the bank of the Chicago River with a bucket of bait and a quiver of fishing rods. He's a regular here.

His gear is all top-of-the-line equipment: three graphite rods and spinning reels on a tripod stand, and attached to the lines is a radio-alarm system that

Carp fisherman Paul Pezalla sits on a mooring button on the south bank of the Chicago River, west of the Columbus Drive Bridge. Chicago River Mile Marker 326.5, June 2002.

functions as a high-tech bobber to alert him to a strike. It's all to help him catch the only fish that interests him: the carp.

Pezalla is a shy, smiley man with a head of Mark Twain hair and mischievous eyes that come alive when he talks about his pastime. The much maligned carp, he tells you, is an underrated sports fish that regularly hits twenty or thirty pounds. That size of fish has a lot of fight.

The numbers seem to bear him out. The Chicago River has an international reputation as a carp fishery, and each year the Chicago Carp Classic attracts hundreds of fishermen from Europe and Africa.

There's a lot of money spent on the sport, too.

"It's a little embarrassing to say," he says, pointing to his gear. "But there's probably $1,500 worth of stuff when I'm all rigged out like this."

Pezalla is like a lot of carp fishermen in that he takes his fishing seriously but with a sense of humor and a sort of bemused detachment.

"Let's be honest about it," Pezalla says with a nod and a wink. "It is pretty crazy."

It turns out the support group for people like Pezalla is quite large. He is a founding member and president of a local carp-fishing club, and he started a business that caters to other carp fishermen. It's called Wacker Baits, named after the street, not any sort of mental condition.

He sells his own brand of carp bait on the Internet. Today he's trying out a new recipe of bait, digging through a bucket of yellow-gray meal.

"Everyone has his own recipe. Let's see, this one's got some field corn, some ground-up nuts and seeds. And it's got some hemp oil in it," he says, flashing those mischievous eyes. "It works."

There's another carp fisherman with a line in the water not far down the wall. Jim Monahan, a chef by trade, is taking a break from work to test his luck. He knew Pezalla would be here. They've been fishing buddies for a while, and he is cofounder of the carp-fishing club.

"Carp are fun to catch. People don't realize what a great widespread, wonderful fish you have in the carp," Monahan says, cheerfully waxing on. "It's also one of the most consumed fish in the world. I wouldn't suggest you eat a carp out of the Chicago River, but you can go to Chinatown right now and buy live fresh carp."

But these guys, Monahan and Pezalla, aren't interested in carp as a food source. It's all for the fun of it, the laughs, and it's strictly catch-and-release.

"I do catch-and-release mostly for philosophical reasons," Pezalla says, and then he has a moment of truth and adds, "But I've tasted them, and they're not all that good."

Summer is prime carp-fishing season. Spawning takes place in late May and throughout June, so around then it's not unusual at this spot to catch between twenty and thirty fish averaging about ten pounds each. Both of these carp anglers have caught individual fish close to thirty pounds. But that just whetted their appetite for more.

It's not surprising that carp fishing in the Chicago River is improving. The water is cleaner than it's been in decades, and other species are showing up in greater numbers, too.

Suddenly, there's a commotion about twenty yards down the bank. Some fisherman, not a carp angler, has caught something odd. A small crowd has gathered. "What is it?" someone asks. A net comes out; the fish is lifted.

"Oh, my God, it's a coho!"

Monahan shakes his head. A coho? In the Chicago River? "If we get a salmon in here, I'll just . . ." his voice trails off. "I've never seen anything like that in this river."

Sure enough, it's a coho salmon. It's small, to be sure, about fourteen inches long. But it's a salmon, all right, the first any of these anglers has seen in the Chicago River.

"Throw it back!" Monahan suddenly shouts. He turns to Pezalla and says in mock seriousness, "This used to be such a nice spot. Now we're getting all these garbage fish in here."

Under the Ashland Avenue Bridge

Grease on the Palms

Chicago has more movable bridges than any other city in the world. There used to be more than fifty of them, but today their number is closer to forty-five. As one heads into the city from Chicago's wintering marinas on the South Branch, the first one encountered is at Ashland Avenue.

This is what is known as a bascule bridge, a style that was first engineered more than 100 years ago. It's the same design that's used in the Tower Bridge in London, but there are so many of these bridges in this city that it's become known here and elsewhere as the "Chicago-style" bridge.

The bridges operate on the counterweight principle. When their massive gears are set in motion, gravity takes over, and hundreds of tons of balanced concrete swing on rocker arms into the pits below, making the two bridge decks rise and separate. When the bridges are open at the same time along a straight stretch of the river, they look like so many palms raised, but not quite pressed, in prayer.

Machinist Paul Arteaga is conducting seasonal maintenance on the bridge, getting ready for the big boating days of spring. He and his crewmates are checking the power and the working parts. They grease the gears and pump seepage out of the hole. If enough water collects down there, the counterweights can't swing into the pits and the bridge won't rise. So it has to be pumped out periodically.

When they descend into the belly of a bridge, they don't know what they'll find down there—old homeless people, kids on the lam, trash, the nests of vermin. Sometimes the maintenance crews smell it before they see it.

They found a dead man once. A suicide victim, presumably. He was hanging from a beam overhead, a poor old soul that no amount of prayerful palms could save.

City of Chicago machinist Paul Arteaga beneath the south side of the Ashland Avenue Bridge. Behind Arteaga are the gears that raise and lower the bascule (French for "seesaw") bridge. Chicago River Mile Marker 321.6, April 2002.

Sailboat Exodus

Crowley's Yacht Yard, Chicago River

Each spring, the sailboats at Crowley's Yacht Yard begin their exodus to the lake. From late March through June, more than 700 sailboats will glide up the Chicago River, wending their way through the city to the harbor and Lake Michigan.

It's a migration as regular as a smelt run or the flocking of geese.

The boats have spent the winter at the yacht yard, way down the river and far from the unforgiving squalls of the icy Great Lake. They put in at Crowley's for the season, to make repairs and clean up, get fitted with new gear, receive new paint, and stay out of the water. In the spring, they head back to the lake. Along the way, more than two dozen bridges will open, one after the other, to let them pass in packs.

There are twenty bascule bridges ringing the Loop alone—from Harrison Street to Lake Shore Drive. Most of them were built between 1910 and 1930, but the youngest, the Columbus Drive Bridge, was built in 1982. And there are a handful of lift bridges, too.

In order to minimize the disruption of street traffic, the exodus is done in shifts on a schedule established by the city's transportation department. The boats, most of them fixed-mast vessels, are sent out in packs of about twenty-five, and after they have passed through an opened bridge, the city's bridge-tending crew will close the bridge and drive to the next one. With this leap-frog method, the boats eventually make it to the lake.

Crowley's Yacht Yard, located off Archer Avenue near Halsted, is the largest marina in Chicago for this size and style of craft—primarily sailboats of at least thirty-five feet. But within that class of boat, there are all kinds of variations.

Grant Crowley, who has owned the yard for more than twenty-five years, has seen a lot of variations of boats and people at the marina.

"We have owners who operate on a $4,000 annual budget, and we have guys who spend $400,000," he says. "We have guys who ride the bus down here to work on their own boats, and we have one guy who shows up in a limousine and gives us his order and tells us what he wants done to his boat."

The water is the common denominator here, and the love of sailing.

Every spring, those people—the rich sailors and the public-transportation types—find themselves floating next to each other in the Chicago River, waiting for a bridge to open. They might see each other out on the big water this summer. And next fall, when they head back up the Chicago River to Crowley's, they might end up together again, going the other way, waiting for that same bridge to open.

The view downstream of the Chicago River between Crowley's Yacht Yard wharf and the Commonwealth Edison's Fisk Station coal wharf. Chicago River Mile Marker 322.5, April 2001.

A nearly hidden plaque just downstream of Lemont marks the completion of Section 10 of the Chicago Sanitary Canal in 1895. The plaque is located on the north side of the canal. Chicago Sanitary and Ship Canal Mile Marker 297.5, June 2003.

Change of Scenery

The Handoff in Lockport

The lift at the Lockport Lock and Dam is the most significant of any lock on the Illinois Waterway, a rise of almost forty feet. But it is more than the elevation that changes here.

From downstream, Lockport is where boats leave the river and enter the canals of Chicago. It is here that the narrower Chicago Sanitary and Ship Canal takes over and leads to the Chicago River, the Cal–Sag, and the lake.

From upstream, Lockport signals the end of the straight, concrete-lined channel. Just downstream from the lock, the Des Plaines River joins the flow and lends its name to the waterway, hinting of slightly wilder things to come.

Deckhand Mike Clark handles the bow guide ropes on an upstream-bound barge as the Lockport Lock, about forty feet deep, begins to fill with water. Chicago Sanitary and Ship Canal Mile Marker 291, November 2000.

From the quiet of his Sawmill Lake blind,
hunter Brent Millinger looks into the predawn
skies of duck season's opening day. Parallel to
River Mile Marker 197.3, October 2000.

CHAPTER FOUR # Adaptations

Five white pelicans, in their gawky-gracefulness, swoop low and slip effortlessly into the water. They are part of a larger flock that has paused here on Anderson Lake.

About thirty more pelicans are lolling about fifteen yards away, just off shore. The five newcomers paddle toward the flock. It's a peaceful scene. And then . . . just as the five reach the others, the water erupts in a mad frenzy, a churning flurry of feathers and fish as they slap at the water and come back with beaks full of shad. It's feeding time.

This is how pelicans hunt. Using their webbed feet to herd small fish, a few birds bring food to the rest of the flock. It's a community effort. Soon five or six different pelicans take flight, circle once, and land about fifteen yards away. Then they begin paddling toward the others, and the cycle repeats.

Little more than a decade ago, this sight would have been improbable on Anderson Lake, a shallow body of water just off the river channel in southern Fulton County. One wouldn't have seen a pelican in the valley. But here they are. And their numbers are increasing.

What brought them here is debated by those who study waterfowl, but the prevailing theory is that the floods of 1993 and 1995 created long-lasting, wide expanses of water, and the birds, which had been using the Mississippi River as a guide on their yearly migrations, spread out and discovered the Illinois. The next year they were back, and soon they had established an auxiliary migratory route.

The white pelican's story is one of adaptation, an echo of the ages-long process of natural selection. The environment changes, and species adjust.

For thousands of years, those changes were gradual, at times glacial in pace. But over the past 100 years, the changes have been sudden, in some cases catastrophic, and virtually every plant and animal species in the Illinois River valley has been affected in some way.

For the pelican, the change was brought about by a natural event, but the most profound changes have resulted from the machinations of man. Not all species adapted as well as the pelican. Not all of them survived.

CATACLYSMIC CHANGES

For thousands of years, the wetlands and backwater lakes worked in concert with the river, taking in water

Pusher Mike Green, cook Leona Miller, and club manager Tim Miller on the rear porch of the Swan Lake Hunting Club. The club is more than 100 years old. Near River Mile Marker 198, November 2001.

during times of flood and slowly releasing it back to the channel. Within that framework, a balance of plant and animal species, including human, had developed.

But the reversal of the Chicago River and the sudden diversion of Lake Michigan water into the valley disrupted that millennia-old ecological balance almost overnight.

Initially, the sudden influx of water had a positive effect on some wildlife because it almost doubled the surface water in the valley. The river widened, marshes and wetlands increased, and more waterfowl habitat was created. Bird populations soared, and fish numbers increased, too.

Many of the old-timers who live along the central and lower stretches of the Illinois recall the days when thousands of ducks filled the evening sky. They reminisce about how the backwater lakes were deep and clear, and anyone with a net and a boat could catch enough fish to fill a market.

Indeed, the Illinois River was at one time one of the nation's top fisheries with a production that rivaled that from the Columbia River in Oregon. In 1908, about 2,000 commercial fishermen were working the lower 200 miles of the river, and their annual harvest topped 24 million pounds of fish. That was 10 percent of the total U.S. catch of freshwater fish.

What the old-timers remember, however, was an anomaly, a scene created by the diversion of lake water, and it would not last. Beneath the surface, the river was in shock.

Wastewater from Chicago, both human and industrial, was polluting more and more of the river. From upstream on down, fish kills became common. By the 1920s, the area north of Starved Rock was virtually devoid of aquatic life. And from the river bottom on up, sedimentation was taking its toll, too.

Along with the diversion of Lake Michigan water and the great Chicago flush came an aggressive levee-building initiative intended to protect farmland and towns from a river that now had a lot more water in it. But those levees had a collaring effect on the river that cut it off from the habitat-rich floodplain.

Exacerbating the problem was a systematic drive to drain marshes and wetlands for row crop cultivation. More than forty organized drainage and levee districts were developed between 1902 and 1929 to drain bottomland lakes for farming.

Other agricultural practices clashed with the interests of river ecology, too, primarily the misapplication of chemicals—fertilizer and pesticides. Nitrogen runoff has been linked to algae blooms in the river, which block sunlight from reaching the river floor where aquatic plants are nurtured. And the problem is not simply a local one. Chemicals applied on Illinois farms have been traced down the river system to the Gulf of Mexico; there, a huge "dead zone," an oxygen-starved area, has developed, where no fish can survive.

The net result of these bottomland- and river-altering events and the increasingly intensive land-management practices of the twentieth century was the creation of a new paradigm for the Illinois River. Habitat was destroyed, and dozens of animal and plant species were eliminated or driven out. No species was left unaffected, not even man.

LEARNING TO LIVE TOGETHER

It can be argued that when the Chicago River was grafted onto the Des Plaines River and the valve was opened to the lake, the Illinois River ceased to be a river. It had become a canal, with regulated water levels and volume. Some contend that the locks and dams had transformed the river into a series of reservoirs and sediment traps.

"Except when it floods, this isn't a river at all," says Larry Rice, the site superintendent at the Marshall State Fish and Wildlife Area. The river in Marshall County is part of the Peoria pool. It lies between the dams at Peoria and Starved Rock. "It's all impounded water." The point is that the collared waterway is not able to breathe and pulse as a natural river anymore.

But even without the engineered changes in the twentieth century, the very crush of civilization had profoundly affected the valley. Much of the wildlife—bison, fox, wolf, bear, among others—that inhabited the region with the Native Americans and at the time of Marquette and Jolliet is gone today and not because of any change to the hydrology of the river.

It is estimated, for instance, that when the Euro-Americans began pouring into the region in the early nineteenth century, there were 10 million to 40 million beaver in the Illinois and upper Mississippi basin. But trapping and the destruction of habitat virtually wiped them out.

Obviously, some species were better suited to the new paradigm than others. Patrolmen with the Chicago Police Marine Unit say they occasionally see coyotes along the Chicago River just outside the Loop. The speculation is that the animals follow the rail lines into the city, where they hunt mice and rats and other urban prey.

The survival story of coyotes, mice, and rats might not sound like a wildlife success story to some, but it is an example of how species learn to adapt to a changing environment, finding new patterns even in the industrial grid of humans.

A similar story belongs to the bald eagle. Once an endangered species and a rare sight along the Illinois River, the eagle has rebounded impressively. The birds winter in Illinois along the river, coming south from nesting areas in Canada and the northern Great Lakes after their fishing holes freeze. The eagles have settled into a symbiotic relationship with humans, roosting in cold months near dams, where there is always open water, and they sometimes follow towboats, watching for fish to be churned up in the wake.

"They'll fly right next to the boat, so close you can almost reach out and touch them," says Luke Moore, a riverboat captain and vice president of Western Kentucky Navigation. "There'll be groups of them, ten, twelve. . . . It's really something to see."

There are other examples of wildlife adapting to the presence of human beings. Geese and other migratory waterfowl, for instance, have learned that the lakes don't always freeze at the power plants that dot the riverbank near Morris and Joliet. Sauger are plentiful in the Peoria pool, and the spring spawn is prime time for sport fishermen in the moving water just downstream from the dam at Starved Rock.

THE INVADERS

Also adapting to the changing environment are less desirable species. These are the invasive, nonindigenous species that found their way into the waterway from foreign waters.

The round goby, a small fish native to Europe, entered the waterway as the zebra mussel did, through the ballast of foreign vessels that use the Great Lakes. The problem with the bottom-feeding goby is that it is an aggressive and prolific fish that gobbles the food source for other fish and stirs up sediment. If left unchecked, it will out-eat all other species.

To stop the goby from spreading downstream, the U.S. Army Corps of Engineers and the Illinois Department of Natural Resources installed an electronic barrier on the Chicago Sanitary and Ship Canal near Romeoville. The barrier is an electrical shield, emitting a low-powered current through the water, which stops fish migration either upstream or downstream. So far, it has been effective.

The remains of a tree reconfigured by a beaver, just south and downstream of the Interstate 474/Shade-Lohman Bridge and upstream from the Peoria Lock and Dam. The gauge shown on the downstream protective cell indicates the number of feet from the water line to the bottom of the bridge for overhead clearance consideration. River Mile Marker 157.8, November 2001.

A different invader is coming from the south: the Asian carp. Imported in the 1970s to control vegetation in aquaculture ponds in Arkansas, these large fish escaped into the waterway during a flood in the 1990s, and in one decade they have spread at an alarming rate. They moved up the Mississippi River and into the Illinois and are out-competing the commercially harvested buffalo, catfish, and other carp species.

Like the round goby, the Asian carp stir up the bottom and cloud the water, making it difficult for aquatic plants to grow. And unlike other species of freshwater fish, they are well suited to survive in a sediment-laden waterway. The barrier that keeps the goby out of the Illinois River also helps prevent the Asian carp from invading the Great Lakes, but there is nothing stopping the carp from moving up and down the rest of the Illinois.

The most notorious invasive species is the zebra mussel. The tiny mollusks attach themselves to hard surfaces such as intake pipes, concrete walls, boat hulls, working machinery, and even the shells of native mussels. If unchecked, they can plug pipes, halt machinery, and kill other mussels.

Although the zebra mussel caused alarm along the waterway in the last few decades of the twentieth century, its population seems to have leveled off, and some ecologists speculate that the Illinois River is proving to be an inhospitable environment because of its relative warmth and the amount of suspended silt in the water.

THE HUMAN RESPONSE

The Illinois River, before the diversion of Lake Michigan water, was prime habitat for a different type of mussel, the freshwater variety. They thrived in the gravel beds and clean current of the Illinois, and they were some of the first natural resources to be harvested from the river. Native Americans used the meat as a food source and the shells for tools.

In the nineteenth century, the shells were found to be a plentiful and adaptable raw material in the manufacture of buttons. They were easy to cut and were durable. As the button became an almost ubiquitous fashion accessory, demand for shells skyrocketed. From Peoria south, the shelling industry boomed, reaching its zenith around 1920, when more than 2,600 commercial mussel fishermen were operating on the Illinois River.

While the harvest was on, entire families would camp for weeks on the riverbank, cleaning their catch in big boil vats, preparing the shells for the rough-cut factories. Towns in the lower valley had factories where round "button blanks" were cut from the shells, and those blanks were then shipped to finishing factories, primarily out-of-state operations.

For almost half a century, until the 1940s, the harvest of mussels and the making of buttons helped power the economy in places like Beardstown, Meredosia, Naples, Kampsville, and a town that earned its name from the industry: Pearl.

The remnants of Pearl's old button factory lie dormant on the way out of town, a roofless brick structure overgrown with weeds and trees. It is a reminder of a once-vibrant economy and a way of life.

The bottom fell out of the button-shell industry in the 1940s with the advent of plastics and the invention of the zipper. Meanwhile, the mussels themselves were taking a double hit: they had been overharvested for years, and the environment was turning against them as industrial and human waste was polluting the water and siltation was slowly burying their habitat.

The freshwater mussel has become rare in the Illinois River. The Department of Natural Resources estimates that half of the mussel species that inhabited the river 100 years ago are gone today. And many of those that survive are either threatened or endangered.

What happened to the freshwater mussel, and to the economy and subculture it spawned, illustrates an obvious and grim lesson: that diverse economies survive. The map of the Illinois River is peppered with the remains of stunted communities and ghost towns. Not all were shelling towns. Quarries and mines played out; the commercial fishing industry fell precipitously; canals closed and railroads bypassed; and bridges were built elsewhere. Any of these factors could kill a fragile local economy.

When demand for a particular product drops, one either finds something else to sell or suffers. When a town places all of its economic eggs into one basket, it sometimes has to eat those eggs itself. And when they're gone, it goes hungry.

From Chicago to Grafton, towns no longer rely entirely on the river harvest for their survival. No longer do they look for their income to arrive by boat. Many have invited new businesses—grain elevators, casinos, packing plants, and prisons. They are adjusting to change in a new world.

Many, particularly those downstream from the Big Bend and Depue, are turning to recreation—seasonal hunting, sport fishing, and boating, using the river for something different. Some are tapping the well of history, promoting their museums and heritage. They're hosting festivals and selling antiques and pieces of their past, trading with tourists. They are adapting.

Again, the town of Pearl provides an example. A new business has moved in, attracted by the availability of another cheap and plentiful raw material: the Asian carp, the fish no one wanted.

The company, Big River Fish, is a processing operation that smokes the fish to make an array of products used in soups, kosher foods, and imitation salmon. Smoked carp is favored by some Asian and African cultures, and the market is huge. Co-owners Lisa McKee and Rick Smith processed over a million pounds of fish in the business's first full year in Pearl.

The business found a silver lining in the cloudy water stirred up by an invasive species. And Big River Fish buys its carp from dozens of fishermen and employs more than a dozen people in a town that hasn't seen prosperity for nearly seventy years.

Stephen Havera, director of the Bellrose Waterfowl Research Center at the Forbes Biological Station, in a tower that overlooks the Chautauqua National Wildlife Refuge, north of Havana. Parallel to River Mile Marker 124, July 2003.

The future of humans in the Illinois River valley is, in many ways, self-determinant. We possess the ability to change our environment, something the mussel, bison, and beaver never had. And how we respond to changes in the environment and the economy will determine our own survival rate.

Voice in the Wilderness

In the Wind at Chautauqua

Stephen P. Havera was working at home several years ago, writing another chapter for his book, *Waterfowl of Illinois*, but the rumble and roar outside dragged his concentration down the road. That's were the earth-moving equipment was tearing out trees along a creek and draining a wetland.

The irony is that he was writing the chapter on waterfowl habitat, the very thing that bulldozer was destroying.

Havera is the director of the Bellrose Waterfowl Research Center at the Forbes Biological Station near Havana, and the book he eventually finished in 1999 is considered the definitive work on the subject. At 620 pages long, it is a largely scientific tome with charts and tables of statistics, but it is laced with history, and its conclusion states the case for habitat preservation.

Havera has spent most of his adult life fighting to save what's left of the migratory bird habitat in the Illinois River valley. He's done the fieldwork; he's given speeches; he's written articles and reports; he's led workshops. And yet, he says, his voice seems to fall on deaf ears sometimes.

The landowner with the heavy equipment was tearing out a row of willows along the creek and preparing to drain a forty- to fifty-acre wetland on his property. Havera shakes his head as he tells that story.

"And then in the fall, he pumps water on it and thinks he's going to hunt ducks! He's one of the biggest hunters around, and he doesn't have a clue."

He and others have devoted years and vast amounts of energy trying to educate the public about the needs of waterfowl and to convince people that the welfare of migratory birds is important. The health of the river, he contends, is vital to the survival of not only waterfowl species but also our own. To Havera, the equation is simple: If we don't preserve habitat, we won't have wildlife; if we don't have wildlife, we lose a part of our heritage.

Havera is walking along the edge of Quiver Lake on the Chautauqua National Wildlife Refuge and pauses to look out at a flight of ducks. It is early autumn, and the birds will be flocking back in large numbers pretty soon.

"In a couple of weeks, you come out here at dawn or dusk, and it's something to see," he says.

He's seen thousands of those dawns and dusks, and every one has been different. There is an intangible link between humankind and the wild. People are bound to it in ways that words can't explain, but this is Havera's passion, and he never quits trying to articulate it to others.

"We need this," he says, cocking his head out toward the lake. "We need it as much as the waterfowl do."

Havera joined the Forbes Biological Station more than thirty years ago. The lab, the oldest wildlife field station in North America, has been conducting research since its founding by Stephen A. Forbes in 1894. Operated now by the Natural History Survey, a division of the Illinois Department of Natural Resources, the lab conducts research into all manner of aquatic life, both plant and animal. There are about ten full-time scientists—waterfowl and wildlife specialists—stationed here. Together, they and their staffs monitor about 800 miles of rivers, including the Illinois and the Mississippi, and dozens of lakes. They study fish, waterfowl, songbirds, plants, and water quality.

Their studies have proved repeatedly how vital the backwater areas and wetlands are to a healthy river ecology. Wetlands support a wide variety of plants, which provide food and habitat for all sorts of wildlife, not only birds. Fish need these areas for spawning and to feed. These wetlands also act as filters for drainage, trapping sediment and agricultural chemicals before they reach the channel.

Migratory birds need these places to nest and rest—they need places like Banner Marsh, the sloughs and swamps at the mouths of creeks, and the backwater lakes like Anderson, Spring, Rice, and Chautauqua. Some of these birds travel thousands of miles in a season, and without adequate

habitat, they would crowd into more confined areas, where they would be susceptible to disease.

For almost 100 years, the wetlands of the valley were drained and leveed off, the banks of the river and its creeks were cleared of vegetation, and irreparable harm was done to the river system.

"We have changed the landscape more in fifty years than it would take nature a thousand years to do on her own," he says.

Yet lately, there has been progress toward wetland restoration.

Congress passed the Clean Water Act in the 1970s, and there are federal and state initiatives that encourage farmers to take marginal acreage out of production in order to create buffer zones between their fields and the river. The state's department of transportation takes part in a wetlands mitigation program, which is intended to replace any wetlands affected by capital improvement projects. Federal and state refuge lands have expanded, too.

In the private sector, duck clubs provide valuable habitat for waterfowl. And the Nature Conservancy has two large tracts of land in the lower reaches of the river that it hopes to use as a breeding ground to improve the biodiversity of the river system. Some large farming groups, like the Illinois Corn Growers Association, have also invested in a cleaner river, as has the barge industry.

"I think we've made inroads there," Havera says. "I think we've increased public knowledge and interest in restoring wetlands in the floodplain. And it's better to do it now, or we might not get the chance to do it."

Any optimism he might harbor, however, is anchored by the weight of reality. It is important to recognize that there will always be floodplain farming, he says, that the river is needed for agriculture, for the barge industry, for lots of things that don't quack and honk.

"You hear people talk about reclaiming the entire floodplain," he says. "Isn't going to happen. There are people who want the locks and dams torn out. Isn't going to happen."

High-production agriculture is here to stay, too, he says.

"It's just like the navigation of the river. Let's face it; there's nothing anyone can do to stop that. Heck, it's just a very economical way to ship corn and coal. We'll just have to deal with it."

His point—and it is one made by others, the Corn Growers Association and commercial barge haulers included—is that as long as everyone in the valley has to share the same resources, they should work more closely with each other. They need deep water for navigation and cheap transportation. They need a place to send runoff; they need protection from floods. And they need a place to go watch the birds.

Havera states the case in the final chapter of his book: "A reasonable balance needs to be established between economic growth and the preservation of our natural resources."

But for all the successes achieved at the govern-

mental and organizational levels, Havera says, the matter still comes down to the individual. So much depends on the farmer who chooses to buy in to the set-aside program, to not plow to the stream bank. So much depends on the rural homeowner who decides to get his mower out in June and take out prime rabbit habitat, thinking he's just cutting weeds. So much depends on the landowner who decides to bulldoze the trees along his creek.

Another flight of ducks passes low across the water of Lake Chautauqua. Havera watches them pass and then points out a different group of birds closer to shore.

About a dozen cormorants and pelicans are double-teaming a school of shad, working together as they feed. Overhead, a flock of gulls circles, waiting for the leftovers. In a tree about forty yards away, an eagle keeps an eye on the proceedings.

There are several different types of birds working in concert with each other for a common purpose, he says. "If only we humans could work together like that."

At this moment, Havera is smiling. He does not look to be a scientist burdened with worry, with a voice drowned in the wake of an earthmover. Out here at the refuge, he is a man in his element. And he tunes his ears to the voices in the air, to the call of those ducks in the wild.

"Sometimes, you just have to come out here and listen to the birds," he says. "They'll tell you what you need to do."

Tapping the Seed Bank

The Challenge at Emiquon

A drive through downtown Havana can trigger more than a few memories for Doug Blodgett.

"I remember they used to have poker games in the back room of this place," he says, gesturing to a little brick building. "But they never let us kids in back there."

He laughs at the thought, and he smiles at some other untold remembrance. His mind is tracing the steps of the ten-year-old boy he used to be. He wheels left and points out a shop that used to be a grocery. And there was a restaurant over here, and this place was a tavern. It still is.

Blodgett is the site director for the Great Rivers Area for the Nature Conservancy, a private environmental organization with offices near Havana, and he pinches himself every now and then to convince himself that it's all real. He has his dream job, living near his hometown. And his life's work is inexorably linked to the river he once threw rocks into as a kid.

"When I was growing up here, all my friends couldn't wait to get out," he says. "But I was thinking, why? I liked it here."

It's easy to understand why someone coming of age would want to leave Havana. Jobs are scarce; opportunities are thin. And in some respects, the town is in the proverbial middle of nowhere—the nearest big city, Peoria, is thirty miles away; Springfield is fifty miles in the other direction. But there is a rich cultural heritage here, and the region is blessed with a broad diversity of wildlife and landscape, including the river. As such, it has become the focal point for a number of interests.

Anthropologists and archaeologists have long studied this area because it had nourished various Native American cultures for thousands of years. Dickson Mounds State Museum and National Historic Site is just across the river.

Havana is home to many governmental and nongovernmental agencies. Maintaining offices here are the U.S. Fish and Wildlife Service, the Illinois Department of Natural Resources, and the state's Natural History Survey. And located nearby are a state forest, a state nursery, a state fish hatchery, and Chautauqua National Wildlife Refuge.

This might be the middle of nowhere, but in some respects it is the center of everything.

On the other side of the river from the city of Havana, just north of the Spoon River, is the Nature Conservancy's Emiquon project. This is the linchpin of the organization's efforts to preserve and restore the Illinois River's ecology, and this is where Blodgett works.

At first look, Emiquon seems to be little more than a very large, flat farm, row after row of crops sloping up from the river to the bluffs. Blodgett, however, sees something different. He sees this piece of land as it must have been 100 years ago, and he sees a future that doesn't look anything like the vast field of corn it is now.

Before 1920, two large backwater lakes occupied this area, but they were drained and the river was muscled behind a levee. The place has been in row crops ever since.

In 2000, the Nature Conservancy purchased the 7,000-acre Wilder Farms, the largest farm in the state, and renamed it Emiquon to correspond to the Algonquin Indian name for the region. The land is still leased to farmers as the transition is made to the new owner's vision.

The Emiquon project is multi-layered. The tract is so large that a number of different ecological systems can be linked. There are upland forest and floodplain habitats waiting to bloom here. It's a bold plan, but if successful, Emiquon would provide habitat for a variety of species, helping to reverse a century-long trend that has diminished the biodiversity of the waterway.

Central to the plan is the concept of reconnecting the land and the Illinois River, to let them feed and fuel each other. That will be the biggest challenge.

"You can't just take the levee out," Blodgett says. "The river would come in, dump sediment, and we'd end up with a mud flat. We have to manage the connection with the river."

Indeed, the Nature Conservancy spearheaded a sandbagging operation and ran its pumps at full tilt during a recent flood. For now, this portion of the floodplain can't handle a flood.

But how do you manage the connection? The answer might lie downstream about forty miles, across the river from Meredosia, at a little plot of land called Spunky Bottoms.

The Nature Conservancy purchased a 1,157-acre bottomland farm in Brown County in late 1997. It had marginal agricultural value because it took a lot of pumping to keep the water out. It was too low and too wet, and the levee was not reliable.

"They wouldn't always get their crop in on time, and sometimes they wouldn't get it out in time," Blodgett says. The Nature Conservancy knew, however, that this poor farmland might make an excellent wetland.

They started managing the property in January 1999. The site's land steward, Tharran Hobson, led an operation that spread hundreds of pounds of prairie grass seed and planted more than 6,000 hardwood trees on the property. By the next spring, the prairie had taken hold, and the area was teeming with wildlife.

"Look at that, see?" Blodgett points to great blue heron standing in a knee-deep pool. He's driving on the levee around Spunky Bottoms, and he's like a kid, interrupting himself to call attention to a flight of mallards or a muskrat house.

"When we bought this, it was ditch-to-ditch ag. We reduced the amount of pumping and tried to stabilize the flow of water. What we'd like to do is manage the water, cover the area in the spring, and let it dry out in the summer."

If allowed to stay on that schedule, which is the natural cycle of this river, the floodplain would rebound. That's the theory, and it is proving accurate at Spunky Bottoms.

The Nature Conservancy Great Rivers Area director Doug Blodgett, on the conservancy's boat, downstream of U.S. Route 136/Scott Lucas Bridge at Havana. River Mile Marker 119.6, July 2003.

The Nature Conservancy's Tharran Hobson and Jo Skoglund at an old pump station, which was used to drain the fields for agricultural purposes during the farming operation days at what is now Emiquon. A group of students can be seen in the background listening to director Doug Blodgett explain the project. River Mile Marker 122.7, September 2001.

"It is really something," Blodgett says, "to see a cornfield one day and come back in a year and see 16,000 waterfowl out there. I wish everybody could see this."

A group of Canada geese prepares to land, and another great blue heron lifts out of the cattails and lumbers into flight. Three whitetail deer graze on the new prairie.

Spunky Bottoms is providing valuable lessons for the Nature Conservancy, lessons that might be applied to Emiquon. And from Emiquon?

"The Nature Conservancy is thinking about the entire river as a project," Blodgett says. "The Illinois River is one of only three large-floodplain rivers in the nation considered to have a chance of recovering. We like to think large, and you can't focus on one little area; you have to focus on the entire watershed if you want to solve the problem."

Blodgett points across Spunky Bottoms to some of the plants that are thriving along the standing water: cattail, water lily, and the Illinois lotus.

"We knew there was a viable seed bank out there," he says. "I wish we could take credit for that, but those seeds were out there since the 1920s, since before the levees."

Those seeds might have tried to germinate each spring, only to be beaten back by plows and pesticides. Lying dormant for eighty years, those plants were just waiting for the right conditions, quietly holding to a biological memory of life, of roots and blooms. Life remembers.

Perhaps memory is not the right word, but something links the present to the past in nature. Some migratory birds are guided by it, whatever it is, following routes flown by their ancestors even though they are yearlings and flying alone. Somehow they know which way to go.

And man can be guided by it, too, perhaps more tangibly. He can recall the times he raced down the streets of his hometown, and how as a ten-year-old he would throw rocks from the levee into the water just to see the sun's reflection bounce off in a thousand directions.

Watching the River Flow

On a Porch in Bedford

Bedford—population 28—isn't the kind of town you pass while on the way to somewhere else, unless you're headed to Montezuma—population 14.

And Bedford isn't exactly a destination location, either. There just isn't much here. And it could very well be that the best spot in town is the yard swing at Charlie and Karen Pearson's house.

They were out there one day shortly after the high water of 2002, just swinging slowly and watching the world drift by their home in slow motion. They said hello and shared their view.

The town of Bedford is set on a narrow strip of land between the river and the soaring bluffs. It's barely more than a clump of houses, some of them seasonal

The U.S. and Illinois flags fly next to a former home site that tells river travelers they are in Bedford in Pike County, one of the two smallest communities on the river. Montezuma, slightly smaller with fourteen people, is north approximately one mile. Both are located on the west side of the Illinois River; the view is to the southeast. McEvers Island, which is the far shore, makes up the east side of the river practically from one town to the other. River Mile Marker 48.6, July 2000.

residences, and many of them are up on stilts. Not that that's always made a difference. The river has a habit of making a mess out of a dream or two.

One blacktop passes through town, and it is still covered with mud in places. A few weeks ago, it had been under several feet of water because the river is not leveed here. The main channel is in plain view, and every boat plying these waters has to pass in front of the Pearson house. When it floods, you can see the waters rise from a week away.

The Pearson house is one of the few not on stilts, but it rests on a piece of high ground on the edge of town, and it's probably the only house in these parts

with enough self-confidence to have a basement. The land here has been in the Pearson family for at least three generations.

"I lived here all my life," Charlie says, getting right to the point. "I wouldn't live nowhere else. My dad grew up here; this was my grandfather's farm."

The river has occasionally tried to persuade him to move.

This latest flood brought water halfway up the mailbox post and into the driveway. It rousted most of the other homeowners, who had to cut across the Pearson ground behind the barn to reach the black-top out of town.

But the high water of 2002 was nothing compared to the floods of 1993 and 1995.

"Some of the folks here jacked their houses up higher," he says. "They shot the grade off the levee across the river and built them three inches higher than that. Well, here comes the flood and they sand-bag the levee over there. No one figured on them sandbagging that levee."

Those who raised their houses discovered they didn't raise them high enough. And water was up to the base of the Pearson house. They used sandbags to keep it out of the basement, but they didn't leave the house.

"One night I was sitting inside, and I heard an awful roar," Charlie says, relating how a big chunk of the levee had given way across the river and millions of gallons of floodwater flowed into the fields of south Scott County. Within hours, the water around the Pearson house dropped about ten feet. The relief, however, was temporary.

Once the floodwater had rushed through the breach and filled the fields on the other side, the backwash began. Soon the water level was back where it had been, bringing with it an even worse problem: the stink.

Charlie shakes his head at the memory. "There were corncobs and all kinds of things rotting in that water. The stench was so bad we had to get out. We drove over and slept in our car up on the hill."

But the water never did get into the house. And although it's been close, the Pearsons don't seem too concerned about the possibility of another flood.

"You just don't think about it too much," he says. When you live on the bank of a river, you should expect to get your feet wet. Floods happen. But there are plenty of good things about living here that make it worth taking the risk.

While he talks, a string of barges comes into view, pushed by the *Karla*. She's headed upstream with a short tow. Charlie and Karen watch it pass. Karen had been quietly listening as her husband told his stories, but once the *Karla* is gone, she says, "We have a porch out front that faces the river."

She is a woman of few words, but she chooses them well.

"Sometimes we sit out there at night. We watch the boats," she says. "It's pretty."

Charlie nods. The *Karla* slips out of earshot, and a bird calls. The swing creaks. And imperceptibly, the water drops a little more.

Higher Ground

Solution in Chandlerville

Roy Brown of Chandlerville had had enough. If the water was going to rise again, he decided, his house would, too.

Situated far from the river, the house nevertheless lay in the floodplain of the Sangamon River, a major tributary of the Illinois. When the water rose on the Illinois in 2001, it backed up the Sangamon, and Brown decided to take action.

With help, he jacked up the entire structure and

Roy Brown and his nephew Luke Shores in front of Brown's home during the construction project that will elevate the recently flooded Chandlerville house. Chandlerville is just south of an Illinois River tributary, the Sangamon River, which broke out of its banks to flood Brown's home on West Illini Street. Parallel to River Mile Marker 92, June 2002.

built pillars underneath. Now when floods threaten and his neighbors fret about moving and cleaning up afterward, he can sit in his home and say, "Bring it on."

Johnboat Romance

Hunting with the Joneses

It's always been the three of them: Ron and Vikki and the Illinois River. It's a love triangle, an amicable relationship, mutually acceptable to everyone.

Ron and Vikki Jones are both in their forties, and they've been married for more than twenty years. They've spent most of that time on or around this river—either fishing or hunting or just winging around in the boat.

"We just like doing things together," Vikki says, not quite sure how to explain how they've stayed together so long. They're best friends, and that's all there is to it.

They both grew up in Peoria and live outside of Sparland in Marshall County now. Their house is a testament to their relationship with the river, cluttered with decoys and trophies, gun cases and fishing gear. The boat is always ready, and they never go far without their dogs—a couple of golden retrievers named BB and Auty.

Like so many couples who have been together a long time, Vikki and Ron seem to know what each other is thinking. There are thousands of shared memories, and they know all the same stories and tall tales.

So they will finish each other's sentences sometimes, punctuate each other's comments, fill in the blanks, and polish the embellishments. Their conversation is sometimes a rapid-fire storytelling duet.

Vikki's grandfather once owned a duck hunting club on the river back in the 1940s, and she starts telling a story about her mother, but Ron chimes in and they end up double-teaming it. It goes like this:

"My whole family hunted, but as far as Dad was concerned, women didn't hunt," she says.

And Ron adds: "So they'd never take the women out."

"But my mom was interested, so my grandfather and my dad took her out hunting . . ."

Ron: ". . . but they didn't tell her a thing."

Vikki: "They just gave her a 12-gauge shotgun and said, 'Shoot that duck.' It kicked her out of the boat . . ."

"She landed in the water."

". . . *out* of the boat!"

"And her mom was like five-foot-two, ninety pounds . . ."

". . . she was a petite little thing."

"And it was November . . ."

". . . it was *cold*."

"She never went hunting again . . ."

". . . never went again."

". . . which was the whole purpose of the exercise."

"Exactly."

Dressed in her typical attire for hunting season, avid hunter Vikki Jones from rural Chillicothe poses with her shotgun at the edge of Goose Lake at the Marshall State Fish and Wildlife Area–Sparland Unit. Parallel to River Mile Marker 190.5, October 2001.

Ron had a different view about women hunting, so when he found out Vikki was interested, he took her out and taught her how to hold a gun and stay in the boat at the same time.

"He was working at Cat at the time, second shift," Vikki recalls. "He'd get off of work, sleep for a couple of hours, and go out hunting. Then he'd go back to work."

"I was younger then," Ron cuts in.

"So I figured if I was ever going to see him, I'd better learn how to hunt."

So the courtship took place on the river, and they've been there ever since.

The Joneses are well known along this section of the river. They're on it as often as they can be. They'll fish for sauger when the fish are pooling up by the dam, and they're out in the blind almost every day during hunting season.

They hunt deer, too, and they fish other streams— there's a place up in Wisconsin that they frequent— but their favorite time and place is duck season on the Illinois River.

Ron speaks reverently about the birds that "originate in the Canadian provinces and fly, some of them, all the way to South America. They are beautiful."

Not everyone has the opportunity to do what the Joneses do. Jobs, school, and kids interfere for most people.

"I have a terrible attitude," he says. "I don't want to work. Some people's whole lives are their jobs, but I've quit jobs because they told me I couldn't be off during duck season. I asked my boss, I said, 'How many sunrises do you see a year?' He didn't have an answer."

Jobs are for paying bills, not for building lives around.

"There's nothing like being on the river in the fall," Ron says, "when the leaves are changing and there's a little snap in the air."

Vikki shares that attitude.

"Give me the smell of an outboard motor," she says. "That's heaven."

The Joneses' philosophy holds that life is short and should be enjoyed. There's no telling how many more flights from Canada to South America are left.

So as often as they can be, Vikki and Ron are out on the water somewhere, probably there right now, either calling in a flight of ducks or just watching the birds fly by in the sunset.

The Duck-Blind Drawing

An Alternative to Camp Crusty

It was early morning. September, teal season. Vikki Jones watched Todd Collins push himself through the muck and water just a few yards ahead of her.

They were trying to set a few decoys for the morning hunt.

And as she watched him, she had a sinking feeling. Nope, something didn't feel right about this. Todd twisted around and looked at her.

"I'm stuck," he mouthed. The sediment at the bottom of the lake had sucked him up, swallowing his feet.

Moments like this are priceless. This had all the ingredients of hilarity. And Vikki would have been busting a gut if she hadn't sensed her own predicament: She couldn't move either.

So they stood out there looking at each other for a moment. Vikki and Todd, two otherwise sane people, friends since grade school, stuck in the mud up to their knees in what might as well have been the middle of the lake.

They looked back toward the blind. The two other hunters in the group—Brent Millinger and Vikki's husband, Ron—saw what had happened. They were shaking their heads. "Nope. We're not going out there after you. No way."

So Vikki and Todd had to strip off their waders and belly crawl through the muck and water like reptiles, dragging their waders behind them, all the way back to the blind.

"We were completely covered," Vikki said, "I mean *completely* covered with Illinois River mud."

It's a moment like this that makes it all worthwhile.

There are all kinds of duck hunters along the river, but they can be divided into two basic groups: those who belong to private clubs and those who hunt on public lands.

It costs thousands of dollars just to join some of the private duck hunting clubs. It's an exclusive group, and the members pay for that exclusivity. Some of these guys are high rollers, big shots in the world of business, celebrities. In return, they get to hunt in a place that's groomed for waterfowl hunting. The wetlands are drained at the right time of year, planted in millet and other feed, and pumped full of water just before hunting season. And there's a lodge, where members are fed and entertained. When they want to kill some ducks, a guide will ferry them out to the blind; when they're done, someone is there to clean their birds and guns for them—and, if they want, help them out of their boots.

On the other hand, there are those hunters who stand in line for a chance to hunt on public property, either state or federal land. These people work a little harder for their birds. They have to build their own blinds and train their own dogs. They have their own boats. On hunting day, they get up early in the morning, set their own decoys, hundreds of them, and then try to call in a few birds.

For the most part, these hunters are indigenous. They are river people who either can't afford the private club fees or just don't want to join.

Ron and Vikki Jones, Brent Millinger, and Todd Collins do not belong to a private club. They've been doing it the hard way for more than twenty years. And they've been doing it together.

If people want to hunt ducks on public land in Illinois, they have to participate in the annual duck-blind lottery in July. Each public hunting area in Illinois has just so many designated spots, and hunters are allowed to build blinds at those locations if their names are drawn during the lottery. And every year these four hunting buddies throw their names into the basket at the Marshall State Fish and Wildlife Area south of Lacon.

On the morning of the drawing, there's a rumor of rain, chance of storms. But the sky is clear, and with temperatures in the upper 70s, one couldn't ask for a better day. This is an outdoor affair, and more than 400 people are expected for the 2 p.m. lottery. Rain would not be good.

Scores of trailers are parked in the campground. The vendors' tents are up, and the tailgates are down on dozens of pickup trucks. By 11 a.m., the beers have already been opened, and hangovers from the night before are being acclimated to a new day. Food would help, and the grills are starting to smoke with burgers and brats and hunks of chicken. A storm would tear down the big top of this circus.

At the campsite where the four friends have set up, Brent Millinger casts a seasoned eye to the west. The sky is still clear. He's a quiet man of medium build, down-to-earth.

As a hunter, one would expect him to be gauging the wind, and perhaps he is, but there are other odds to figure, too. Like, what are the chances of getting a duck blind this year?

Todd Collins, Ron Jones, and Brent Millinger are stationed in their duck blind on Sawmill Lake, northeast of Henry, at the so-called Camp Crusty. The water level was extremely low because the wind was blowing off the lake and toward the river. The tracks left between the grounded decoys are from Jones's golden retriever, Auty, who had lumbered through the mud to retrieve the first downed ducks of the season. Parallel to River Mile Marker 197.3, October 2000.

There are twenty-four locations and more than 400 names in the basket. If four hunters work as a team, there are more than 100 teams vying for those blinds: a 1-in-4 shot at best.

But not all the blinds are desirable. Some will be inaccessible because of low water and siltation. Others will be in poor locations—either too far from the refuge or too close to swift water. All of these factors could raise the odds of snagging a good blind to as high as 1-in-10.

Getting a good draw in the lottery would be nice. Millinger, Collins, and the Joneses have had good spots here in the past, but there have been times when the foursome was completely shut out. When that happens, hunting becomes an even bigger challenge, and these four usually end up down at Camp Crusty, the scene of the infamous "I'm stuck in my waders" incident.

By 11:30, a bank of dark clouds has pushed onto the horizon, and a steady breeze has started to move through the campground.

Camp Crusty is the name these four hunters have given to a spot of private land they sometimes lease on Sawmill Lake, just across the river channel from Henry. It's not much of a camp. And it's not much of a lake, either.

When the wind is right and the water is up, Sawmill Lake can be just about perfect. No one's around, and it's easy. But most of the time it's so filled with sediment, it's not much more than a mud flat.

They need a boat with a mud motor to get to the blind, and the dogs can barely swim in it. The decoys, which don't have enough water to become buoyant, list to one side, and if the hunters wade out into the muck to set them straight, they run the risk of getting sucked to the bottom. Just ask Vikki and Todd.

"It's a semisoluble," Ron Jones says. "Not quite water, not quite solid. I've got big feet, so I can almost walk on it."

Jones is indeed a big guy. At six-foot-three, he's difficult to miss, and he's well known in these parts. He's friendly and jovial, and he doesn't let much get in the way of a good time. One would have to try hard to ruffle his feathers.

At the duck-blind drawing, he's not at all anxious about the odds. Someone flashes him a pair of crossed fingers. Good luck, it says. He replies with an exaggerated shrug. Who cares?

"Hey, if we get a blind today, great; if we don't, that's okay. We've never *not* been able to hunt. We'll find something."

The fallback position is always Camp Crusty.

Larry Rice is watching the skies with some concern.

As the site superintendent at the Marshall State Fish and Wildlife Area, Rice is the master of ceremonies at the lottery. He's going to try to make 400 people happy, but that might be difficult if the weather doesn't cooperate.

Rice has a lot of admiration for all of these hunters. They have to work for their ducks. If they are lucky enough to be on a team that gets a spot in today's lottery, their work is just beginning.

"What they win today is the right to build a blind," Rice says. "There are guidelines they have to follow—minimum dimensions, what materials they can use, where it can be built."

The blinds, which have to be inspected and approved by Rice and his crew, are typically big enough for four people to stand side by side under a roof. Most have a cutout and a ramp for the dogs to use. Attached to the blind is another shelter that conceals the boat, and then all of it is covered with natural vegetation.

"Basically it's a plywood box," Rice says, "and they're out there about 1 a.m. checking in and getting ready for first light."

A gust of wind whips up, and Rice strides over to the auctioneer's truck, which will be used during the lottery. It's almost 1 p.m. He doesn't need a storm one hour away from the drawing.

And then he says, almost to himself, "I think we're about to get hit hard."

The clouds are gray and dark and streaming in low across the river now. Vikki, still relaxed and seated in a folding chair at the campsite, glances at the Suburban she and Ron drove here.

She's not too worried about the weather. If it storms, the Suburban will be the refuge for the four of them. Nice and cozy. Wouldn't be the first time.

"We like it that way," she laughs.

Vikki, Ron, Todd, and Brent have spent many hours in close quarters. And when you get down to it, it's the camaraderie and, yes, the shared misery that make it all worthwhile.

God knows it isn't the ducks they bring home.

"Oh, no, don't do the math," Todd says. "You will stop doing it if you do the math. You'd have to add up the boat . . . the decoys . . ."

"The shells . . ." Brent helps him make the tally.

". . . the time . . ."

Unlike Ron and Vikki, whose jobs are flexible enough to allow them to hunt almost every day, Brent and Todd are weekend hunters. What keeps bringing them back is something that's hard to put a finger on.

"It's just a good time," Brent says. "Best breakfast in the world is the one you have out in the blind."

"Yeah," Todd says. "We spend a lot of time retelling stories we've all heard a hundred times, but there's always a new memory that pops out of it. Why we do it? I don't know. All I know is if I don't do it, I'm not happy."

As Todd is waxing philosophical, the clouds are whipping past. But not a drop of rain falls, and the wind settles down to a breeze. Clouds are giving way to sky in the west, and it looks like there's going to be sunshine soon.

It's half an hour before the drawing, and across the campground, standing next to the auctioneer's wagon, Larry Rice is watching the passing clouds. He is smiling.

Using the microphone, Rice gathers everyone around. Announcements are made, the winners of various product raffles are awarded their prizes—shotguns and shells—and several kids are invited up to pull the names out of the basket for the main event: the duck-blind drawing.

Several hundred hunters are crowded around, some with dogs, some with kids. There's an expectant buzz to the air. Ron, Vikki, Todd, and Brent are standing together, just beyond the picnic tables.

One of Rice's officers has been spinning the basket continuously for the past several minutes. The cards bearing the names tumble over each other.

Then, when it's time, he stops and opens the wire door. One of the kids reaches in and pulls out a card, which is passed up to Rice. He wastes no time reading it:

"This is choice number 1, and it's Todd Collins from Peoria."

Todd hears his name. It doesn't quite register. Then he realizes it's his name, and for a moment he can't move. It's as if he's stuck in his waders again.

A hail of cheers from his partners snaps him out of it. They beat the odds; they secured a blind, the best blind in Marshall County.

For one whole year, they won't have to rely on Camp Crusty; they won't have to trudge down there and slog through the mud in the early morning. This year, they can sit in their privileged blind right next to the refuge. And while they're sitting there, before the first shot, they will swap a few stories and laugh.

Someone will mention how the decoys wouldn't sit up straight because the bottom of that lake looked a lot like the top of that lake, and they'll giggle about it.

And then someone else will tell the story about the reluctant retriever who wouldn't jump into that muck, and they'll all laugh again.

And then someone will recall the time Todd and Vikki got stuck, and their laughter will be so loud they could be heard a couple of blinds over.

Wake-Up Call

At the Fish and Wildlife Area

Larry Rice has had mornings like this:

The big bearded man with the red face and bloodshot eyes stabs one of his fat fingers about six inches from Rice's face. The man is saying something, he's shouting actually, something about how Rice can't run a goddamn day care center, much less a wildlife area. He's saying something else about *his* tax dollars and Rice's salary, and Rice has heard it all before, dozens of times from dozens of guys, some of them armed and some of them drunk. Like this one. And Rice wonders if he'll have to arrest this one, but he hopes he doesn't have to.

But Rice, the site superintendent at the Marshall State Fish and Wildlife Area, has had more pleasant mornings:

He's out in the boat. It's quiet. He fills his lungs, deeply, and watches the sky slowly turn pink in the east, as if God had a finger on the dimmer switch in the sky. A flight of teal passes overhead, so close he could almost count the birds' feathers. A sandhill crane calls from the flooded woods to his left. A fish

breaks the surface of the water not ten feet from the boat, creating a ripple that unfolds wider and thinner, and then it's gone.

Most of the time, though, the mornings fall somewhere in between, between tranquillity and madness.

In the thirty years he's been working at Marshall County, Rice has pretty much seen it all. Good days and bad. Good people and bad. Sometimes it's a hunter. Sometimes it's a hunter with an attitude. And sometimes it's a sportsman who knows him and appreciates the job he does.

Rice manages a dwindling staff of less than half a dozen full-time officers. They all work for the Illinois Department of Natural Resources, and together they oversee the 9,000 acres of fish and wildlife habitat in Marshall and Woodford counties. The job keeps them all busy, tending to the needs of campers and fishermen, hunters and wildlife. And don't even talk about Springfield. The state office has needs, too.

"Out here," he says, walking the grounds outside his office just south of Lacon, "you have to be the sheriff, the mayor sometimes, the biologist . . . you have all these hats to wear."

Rice is a tall, thin man in his early fifties. A long, easy gait carries him lightly across the grounds. Not a twig seems to break under his stride. In conversation, he listens intently, asking as many questions as

Site superintendent Larry Rice at Marshall State Fish and Wildlife Area on the eastern side of the Illinois River. Rice is sitting next to a catch pool that was constructed to retain freshwater before it flowed into the river. The project was created and co-sponsored by the state and by Ducks Unlimited. Behind Rice, to the west, is Babb Slough, which provides access to the river's main channel. Parallel to River Mile Marker 183.5, November 2001.

he is asked. He's a straight talker, but he speaks softly, dressing his authority in a smile.

He is a willow of a man with a temperament that allows him to bend when he has to. He's likable. And that serves him well when he's wearing the hat of a public servant.

The Illinois River valley supports a patchwork of state and federal lands devoted—in varying degrees—to public use and wildlife management. There are state parks, national wildlife refuges, fish and wildlife areas, state conservation areas, and state natural areas. Each has a different mission, although the distinction is sometimes hard to see.

"As a fish and wildlife area, we don't have a visitors' center and a playground. Our emphasis is the habitat, as opposed to providing the overall outdoor playground," Rice says. The predominant demographic he and his staff see is the hunter, specifically the duck hunter.

Planning for waterfowl season is a year-round job—hundreds of hunters attend the annual duck-blind drawing in July, for instance—but during the actual hunting season, which runs from October through December, tending to the hunters' needs is a seven-day, all-day operation. "It's pretty intense. Come January, we all take a deep breath."

As site superintendent, Rice lives on the site, which can be good or bad. He doesn't have to pay for housing or fight traffic on the way to work—that's good; but he's never able to leave his work behind at the end of the day—that's bad. During hunting season, that work sometimes follows

him home. Literally. And it can get kind of hairy. Literally.

"Every morning, you're dealing with seventy-five to eighty people with guns, and they're all dressed in paramilitary gear," he says. "Some of them are nursing hangovers and, I don't know, maybe they had a fight with their wife, and they get here in the morning and it's cold and windy, and you open the door and that's what you're looking at."

He shakes his head and smiles as if recalling a particular face.

"If you didn't know these guys, you wouldn't want to meet half of them in a dark alley," he says. "We don't get your typical person hunting here; we don't get any of those 'beautiful people.' We get pretty hardcore river people for the most part."

There's admiration in his voice.

"There are a lot of people around here who know about the river. They know more about it than I do."

The admiration goes both ways, too. During the duck-blind drawing in July, one hunter nodded toward Rice and offered perhaps the highest praise one is likely to hear out here: "That's a good man over there."

Rice has roots in the Chicago suburbs, but he was attracted to the outdoors because it was an escape from the concrete wilderness of the city. When he was younger, he thought he'd be a hunter, but once he landed here after finishing at the University of Arizona, he found that just being outside was enough. That was in 1974, and he's been here ever since.

After thirty years at one job, Rice says he's looking forward to retirement, but he knows it will be difficult to leave.

"When I first saw this area, I was amazed because there is a sense of wildness here. You have backwaters, and when you're out there—I know there's a road over there and another road over there, and there's development all the time up on the bluffs—but when you're out there, you can't hear any of that. You swear, you look out there and it's as if you were in Minnesota or Canada or down in Brazil or something."

But reality isn't too far away. Rice points out that the conservation area he manages is but a thin ribbon of the landscape. Once out of the valley, the land flattens out, and the biodiversity drops off markedly.

"The wildlife along this area is pretty remarkable. I could see more wildlife in one day than I could in a lot of places, in the best national parks in the United States or around the world," he says. "On the average day we can see river otters, wild turkey, great blue heron, bald eagles, beaver, muskrat, you name it. Anything except the big predators comes through here. And now you see white pelicans, cormorants, sandhill cranes, waterfowl of all sorts."

Rice sometimes worries about the future of this area, though, wondering aloud why anyone would want to build a faster highway to Chicago from anywhere, much less Peoria. And he laments the development up on the bluffs and the siltation problem on the river.

That's enough to make a man hunger for retirement. But then he'll hear a crane call from off in the flooded wood, and he'll stop to listen.

"Sometimes me and the other guys around here will be out there on the boat, doing stuff, checking on a levee or a pump, and we have to laugh," he says. "We're seeing more wildlife on our jobs in one day than most people will see in six months. I can't believe we're being paid for this."

A Lesson and a Hunt

Memories in Beardstown

Sometimes when he closes his eyes and allows himself a daydream, Jim Reick will be hunting ducks. He's a boy, and he's in a boat with his father on the backwaters of the Illinois River. You can't get to those backwaters anymore.

It's a different river in his mind. It's a river that flows in a time when the floodplain lakes were deep and clean and the fish were biting. And the ducks . . . ah, the ducks, they could blot out the sun, they were so thick.

Max Reick taught his son everything he knew about living off the river. He taught Jim how to shoot and trap and fish. And he taught him a lot of other things, too.

"He'd point at the axes and say, 'This ax is for chopping, and this ax is for trimming,' and he'd expect you to remember," Jim recalled. "You didn't want him to have to tell you twice."

Beardstown's Jim Reick near Muscooten Bay, the confluence of the Sangamon and Illinois rivers. River Mile Marker 89, March 2002.

Jim Reick grew up along the river at a time when one generation handed survival skills down to the next. Sons were expected to follow their fathers into the woods and down to the water because that's where the food was. The lessons were tough, but so were the times.

Max Reick died in 1955, and Jim, in his seventies now, still misses him.

Sometimes when he closes his eyes, he'll be in the boat waiting for the ducks to come in. His breathing gets short and shallow. His father is close. He can smell the man's sweat and his overcoat. And he waits. And he can feel the man's hand on his shoulder, feel his breath on his neck.

And he can hear his voice, directing, teaching, still steering him through life.

"Get ready," Max whispers, "here they come."

The Eyes of an Old Market Hunter

"Oh, We Had a Time"

The years are winning this battle. He knows it.

Dale Hamm, notorious market hunter, sits at his dining room table and gazes out the picture window at Bath Chute. At eighty-four, his body is rebelling. He's fought cancer and other ailments for a few years.

He hurts. And although his will is strong and his mind is alert, the string that holds his thoughts together seems a little frayed.

He's told some of his stories a hundred times, and each time he tells one, it sounds the same. He'll recall a tale from his book, *The Last of the Market Hunters*, and it comes out almost word for word. But between those tales, his memories sometimes wrestle with time and place. He can't quite get his ducks in a row. But what a pretty bunch of disorderly ducks they are.

Through his window, Hamm appears to be watching the river. But his eyes are not what they used to be, and what he's looking at isn't out the window anymore.

"Oh," he says, lost in the muse, "we had a time."

Dale Hamm and kin have probably killed more ducks than any other family in the Illinois River valley. They didn't always kill those ducks in season, and they didn't always know the meaning of the word "limit." They were poachers and outlaws, and some of their exploits are documented in court reports, game warden files, and Hamm's own book. Others are confined to memories.

Hamm and his brothers would sneak into private hunting clubs and shoot ducks until they were chased out. They'd kill ducks out of season, ice them down, and haul them to markets in Chicago. They knew that what they were doing was illegal, but it didn't make any difference. Hunting and poaching was what they did. They were good at it. And they had fun. At the time, that seemed to be all that mattered.

The Hamms became a family of river people out of necessity.

"My grandpa gave each one of his kids a farm, over at Sheldons Grove," he says. "Then the Depression hit, the levees broke, flooded them out. Destroyed all of the crops. So my dad got out of that, and we hunted and fished for a living from then on. That's how we were raised."

Dale, born in 1916, was barely a teenager then, the second son in a family of six boys and two girls. He followed his father around, learning how to hunt coon and muskrat and mink. They fished for bullheads and catfish, and they killed ducks by the hundreds and ran them like moonshine to markets in Chicago.

They were living in paradise, surrounded by abundant hunting areas. There were Stewart and Anderson lakes, the Sanganois, Snicarte Slough, and Grand Island. It was a way of life and a way of staying alive.

Where Hamm went wrong was when times changed and he didn't. After the Depression and World War II, the nation was moving forward. Times were prosperous; jobs and recreational opportunities were abundant. Limits on taking wild game were being enforced, and there wasn't much room for the old backwoods mentality.

"At the beginning, we had to do it," he says of unlimited duck hunts. "But then we were having so damn much fun out there, we couldn't stop."

He wasn't alone, not by a long shot, but he was perhaps the best known of the bunch. He and his broth-

Dale Hamm sits near the Bath Chute landing for the ferry that takes private club members to their lodge and several lakes on Grand Island, which lies across the chute from the town of Bath. Parallel to River Mile Marker 110.5, July 2000.

ers and friends would sneak onto 5,000-acre Grand Island and kill ducks whenever they could get away with it.

There is a well-established and well-heeled duck club there, but the Hamms knew the island better than any of the members. They would slip in under cover of darkness and get to the ducks before the club members were out in the blinds. The club members knew Hamm and his cohorts were poaching, but they could never catch them. Sometimes the brazen Hamm would shout taunts.

"Oh, we had a time," he laughs. "We did things you won't even dream about doing, just to kill a god-damn duck."

Hamm tells his stories and giggles as if he were telling tales over beers after the hunt. He's mischievous still. And of course telling the tales was half the fun. Still is.

"Fact is, we'd go out and do that," he says, referring to one poaching raid or another, "and then we'd get into the tavern and we'd tell about it. That's the worst thing we could do, but we'd tell about it because we were getting a big kick out of all those tricks we pulled."

He liked stepping over the line, just on the other side of being legal. It was exciting, it was fun, and it was about to catch up to him.

Hamm made his big mistake in the mid-1950s, when he trusted a man he'd been hunting with for a couple of years. The man was a federal agent. Hamm and several of his brothers were busted along with a dozen other hunters from Beardstown to Peoria.

"I sold him about 500 ducks, I think it was, and they got the whole damn tribe," Hamm says, and he allows himself an impish smile, not at all repentant. When Hamm looks back at his life, he sees a few things he shouldn't have done, and one of them was getting caught.

The bust didn't totally put him out of business. He still enjoyed the hunt, legal or otherwise, but he just kept his exploits a little closer to his hunting vest. Times had changed, and he knew it.

"You can't do that any more," he says flatly. "There's nothing wrong with taking a few ducks over the limit, if that's what you want to do. But you're going to have to pay the consequences if you get caught."

Hamm looks tired today. He wears down easily and shuffles when he walks, holding onto the chair and table for support. He bends over to search the contents of a trunk for old newspaper clippings and needs help straightening up.

"I was disabled a couple of years ago," he says, out of breath. The struggle with the newspaper clippings has clearly beaten him down. He hasn't hunted in two years, and that pains him, too.

"I lost sight in this eye completely. And I can't see worth a damn out of the other one."

The years are winning this battle.

He gazes out the window at his dining room table. What he sees out there, one can only guess, but whatever it is, it makes him smile.

Boat Tavern

Dry Docked in a Memory

It's not much to look at. The Boat Tavern is a weathered wooden structure perched on the deck of an old steel barge. No one's going to call it pretty or, worse, quaint. The wiring is a little dubious, and the exterior could use a coat of paint. And it wouldn't hurt to throw a handful of nails at it, either.

The whole thing, the entire boat—if you can still call it that—is set up on concrete pillars that keep it suspended about ten feet off the ground. They keep it out of the water during floods.

An empty beer keg sits outside, and a big plastic bag of empty bottles slumps on the deck like a homeless soul. In the light of day, the place looks like a hangover, like the best man after the bachelor party.

No, it's not much to look at, but it is a curiosity, and the Boat Tavern can be a sight for sore eyes for anyone who's looking for a cold Bud, a warm hello, and a hot time.

*A typical summer's day at the public boat
landing in Bath, looking upstream on the Bath
Chute. Parallel to River Mile Marker 110.5,
August 2001.*

Some travel twenty, thirty, forty miles to get here—by car and by boat—and on a Saturday night, the parking lot is full of pickup trucks and half a dozen boats are pulled onto the bank. Music is pouring out of the jukebox—from Willie and Waylon to "Sink the Bismarck" and Shania Twain—and the joint is jumping.

It's been said that a church is more than the building in which the services are conducted. The same can be said of a tavern. Yet the Boat Tavern's identity is inherently linked to the structure. It's hard to imagine the tavern without the boat itself.

Believe it or not, this thing used to float, taking the eighteen-mile loop around Grand Island, serving drinks to anyone who wanted to take the ride or who tied up and climbed aboard. But that was a dozen leaks and a couple of near-sinkings ago.

It eventually went into permanent dry dock and was hoisted up onto stilts down on the bank of the Bath Chute. It's still called the Boat, even though its cruising days are over. It's a fixture around here, a landmark, and its story is the stuff of legends.

The Boat was the brainchild of Dale Hamm, well-known poacher and backwater bender of laws. He doesn't own it any more, having sold it several times, but he still lives just up the hill.

When he moved to Bath from across the river in 1966, Hamm purchased a floating fish market and the land where it docked, and he immediately transformed that wooden-hulled market into a floating tavern. That was the predecessor of today's Boat.

He took trips around the island, hosted gambling parties, and sold liquor. It wasn't making him rich, but life was good.

"I had thirteen refrigerators in there," he says.

Occasionally someone would come through, fall in love with the place, and offer to buy it. Hamm would oblige, but the new owners invariably would go broke or mismanage it, and it would come back to Hamm.

"Fact is, I couldn't get rid of the thing."

The last guy who bought it ran it into the ground, literally and figuratively, so when Hamm took it over the last time, he pulled it out of the water and put a match to it.

Meanwhile, he found a small steel barge for sale up at the mouth of the Calumet River. It was twenty feet wide and fifty feet long, and it would make a fine hull for another floating tavern. He bought it. Now, all he had to do was get it down to Bath.

So he enlisted the help of Fred Miller, a buddy from Bath who had an old Chevy pickup truck, and they hauled Hamm's johnboat and a 40-horsepower Mercury motor to Chicago.

They pulled the truck onto the deck, strapped the johnboat to the back of the barge, fired up the outboard, and off they went. And they drew a few strange looks.

"One lady came out of an office building there in Chicago, and she said, 'That's the damnedest thing I've ever seen. Wouldn't it be easier to just drive that truck wherever you're going?'"

It took them five days to get to Bath. They pushed that barge 200-some miles, through five locks, with Hamm at the tiller and driving blind.

"I couldn't see nothing back there," he laughs. "But my buddy—he was kind of a screwball anyway—he'd sit up in the truck, and whenever he'd see a buoy or something, he'd turn on the blinker lights of the truck—you know, either right or left—and I'd know which way to go."

They made it, though, and Hamm parked it down on the beach. He slapped up a building on the deck and called it a tavern, the Boat Tavern.

"It was popular as hell," he says. "We had more business than anyone in the country. I took a bunch of lawyers around the island once because I had this gambling deal going. That was one of the first gambling boats on the Illinois River."

Nothing legal, mind you.

The Boat had a fate similar to its predecessor's. Hamm ended up selling it, and it has changed hands several times since. It developed a leak or two or three, took on water, and almost sank. Somewhere along the line, it was retired from active duty and set up on pillars.

Despite the elevation, it's still called the Boat, and sometimes, on a Saturday night when the music's loud and the barkeeper is handing out beers and the patrons are swapping lies and laughing out loud, one would swear the place is rocking. A person can almost hear the slap of water under the deck and feel the roll of the river.

A Wedding Song

In Love along the Bath Chute

The music can be heard all the way down at the boat ramp.

The karaoke orgy is just getting started up the hill—Waylon and Willie are courting a good-hearted woman—and the river laps at the bank and almost keeps the beat.

The concrete slab extends into the river, waiting expectantly for a trailer and a johnboat. One could go anywhere in the world from this spot. Head north and a person would eventually hit Chicago, and from there it's on into the Great Lakes and beyond. The sky's the limit. The other way leads to St. Louis, New Orleans, the Caribbean. Choose a dream and get in line at the boat ramp.

But there won't be a trailer getting its wheels wet today. No, no one wants to miss the event of the year in Bath. This is the day Kenny DeFord and Betty McDaniel are getting married. So bring out the food, tap the keg, and cut the cake. Hook up that microphone and crank up the volume.

"Well, hello!" Kenny drapes his arm over a shoulder and extends a welcome. "Have you found the keg? It's right over there." He points with a finger and a cigarette to the corner of the tent.

Kenny DeFord is a man genuinely concerned for another's welfare, particularly if that person's welfare depends on a cold beer. He is a generous

man, too. He doesn't have an enemy. Same goes for Betty, who might owe her popularity in part to her job. She runs the Boat Tavern, and everyone around here knows you have to get along with the bartender.

She is found setting one of the tables under the tent. She sizes up the layout. "If the cake goes here," she thinks, "then the plates and the serving dishes can go there . . . so many details . . . oh, who's going to care?"

"Hi!" She welcomes another visitor with a smile and a hug. Then she ushers him over to admire the

cake her daughter Jamie created: three separate tiers, each decorated with an elaborate flourish of iced flowers. "Isn't that the prettiest thing you've ever seen?"

A dozen chickens are on the spit and getting brown, and every time the big black cooker is opened, a cloud of roast-smoke wafts into the tent. As the guests arrive, the side dishes start filling the table. This is a potluck affair, so there's a potato salad and a pasta salad, a green salad and a bean salad. One person brings a box of peaches; another drops off a quart of pickles. No one goes hungry today.

Betty and Kenny DeFord, on the west deck of the Boat Tavern in Bath, looking southwest down the Bath Chute at sunset. Parallel to River Mile Marker 110.5, September 2003.

The wedding ceremony itself is still to come, but the karaoke machine has been going on for a little while now. Folks are taking turns at the mike. A long, tall cowboy-type is crooning now. *"Take the ribbon from my hair, shake it loose and let it fall . . ."*

The bride and groom both grew up in Bath, a town tucked into a crook of the Illinois River. It's not actually on the river itself but on a branch of the river called the Bath Chute. Had it been located on the main channel, Bath might have rivaled Havana and Beardstown—the nearest big towns—in size and prominence. But today, it is home to only a few hundred people and a handful of small businesses along Illinois Route 78 and around the town square. And here, the options are rather limited.

Both Kenny and Betty attended the now-shuttered high school in town but at different times. Kenny, fifty-four, is seven years older than Betty. He graduated high school and was immediately drafted into the army, which sent him to Vietnam.

"When I got out, all I wanted to do was right here," he said. "I just wanted to fish and boat and stuff like that."

It didn't all work out the way he had hoped. When he got home from 'Nam, his mother was already dying from cancer. He shuttled her back and forth to Chicago for treatments—twice a week they'd make the trip—and she lingered for years.

It was at the funeral that Betty first took note of the tanned young man with a face etched by pain and endurance. "He seemed so vulnerable," she remembers.

As in any small town, folks hang out and bump into each other, and more and more often, Betty's circle of friends overlapped his. For years they danced around each other, but eventually they met in the middle of that dance floor. They've been together ever since. *" . . . Layin' soft against your skin like the shadow on the wall."*

"When's the ceremony, Kenny?" someone asks.

"Oh, we'll head down there pretty soon," he says.

To say this is an informal occasion is perhaps overstating the matter. The bride is in tennis shoes and shorts. She and Kenny are both wearing ball caps. Their T-shirts are coordinated, ordered especially for this occasion. Their names and the date are on the front, and on the back of his is the word "Captain." Hers reads "First Mate."

The shirts are appropriate. Kenny and Betty are two river people about to exchange vows on the deck of a houseboat.

"We just did everything together on the water," Betty says. "Everywhere we went, the river was with us. It just seemed like getting married on the river made sense."

They courted in a johnboat. He'd pull up to the Boat Tavern, and they'd head out. They'd run up to Havana or out around Grand Island, which lies between the Bath Chute and the main channel. They'd go to Matanzas Beach or anywhere they wanted on the water.

"One time we were heading to some party up in Havana, and Kenny says, 'Let's take the boat.'" Betty tells the story while Kenny listens, grinning. He knows how the tale goes. "Well, we're almost there, and he decides to take a shortcut across this bay. And then he says, 'What's that up ahead?' The water looked funny, and sure enough, the water was so low Kenny runs us up onto a mudflat. Oh, were we ever stuck."

They both got muddy and wet that day, but they also got a story that still makes them laugh after ten years. And that's just the way they are. Take life in stride and don't sweat the small stuff. Those who grow up along the river tend to look at life that way.

"This is one of the greatest waterways there is," Kenny says, cocking his head toward the Bath Chute and the Illinois. "I can run around all day long for twenty dollars in my johnboat. We can go right out here and head to Chicago or St. Louis, turn up the Mississippi and go clear to Minnesota if we wanted to."

Betty nods; that's right. And the cowboy at the microphone finishes his song. " . . . *Help me make it through the night.*"

It's time for the ceremony, and the bride and groom head down to the water. It's not so much a procession as it is a meandering. Mendelssohn has been replaced by the echoing refrains of Willie Nelson and Kris Kristofferson. A few of the guests have brought their beer in plastic cups.

There's a cinderblock on a wood pallet to help them step up to the deck of the houseboat, and Kenny helps Betty up. And pretty soon it's over. They say "I do," pose for a few pictures, and head back up to the party. Done.

The karaoke machine is cranked up again, and the sounds of the party trickle down to the chute, to the boat dock, where all of this began.

One can go anywhere from this spot, anywhere in the world. It seems that a person's options are limited only by the anchor of the imagination, even in Bath. Some people, like Kenny and Betty, do have dreams. They toss those dreams into the boat like a picnic basket and push themselves into the river. And they find what they're looking for.

Yield the Right of Way

Bridge Tender at Shippingsport

Their days are numbered.

The Shippingsport Bridge will be dismantled soon, its parts destined to replace the gears and shafts at a sister bridge downstream. Salvage.

For now, though, the bridge is manned twenty-four hours a day. The bridge tender, that's Charlie Pieskovitch. He's the one who halts traffic on the highway. He's the one who commands all this steel to rise, 450 tons of it. Smooth and swift, the span lifts, and cars and trucks wait in line while barges pass below. It's clear who has the right of way.

A modern bridge will take the place of this seventy-two-year-old structure, and Charlie will have a job here no more. When the progress of man has the right of way, the man himself is easier to replace than the pieces of machines.

Department of Transportation bridge
tender Charlie Pieskovitch looks out the
downstream window of the Shippingsport
Bridge's bridge tender hut. The open-grate
bridge decking can be seen far below.
River Mile Marker 224.7, March 2001.

Finger on the Button

The Demolition

He presses the white button and slides his finger to the red one. He begins the count. "One, two . . ."

Roy Wasielewski is staring at the Shippingsport Bridge from a trailer on the bank. It's a brisk late-winter morning. The old steel structure that spans the Illinois River at La Salle is stark and silent, and the day's first light is a blue-pink scrim behind its skeletal silhouette.

It's quiet in the trailer. Roy thinks about old dogs and coyotes.

". . . five, six . . ."

Roy has worked for almost twenty-seven years on that bridge. He spent his shift in a little shack atop the movable center span, sharing the space with the motors and the gear shaft that whirled and cranked whenever the bridge was raised. This is a lift bridge, one of the few remaining on the waterway, and Roy is a bridge tender.

The shack's gone now, and the old bridge has been prepped for today's ceremony, its final official duty. The center segment is in the up position. Its steel has been partially cut to weaken it. Holes have been drilled and packed with explosives.

It's suffering, and it's just a matter of time now.

". . . nine, ten . . ."

For more than seventy years, the Shippingsport Bridge carried traffic over the waterway between Oglesby and La Salle, between the lowlands on the south bank to the bluffs on the north. At one time it was the only bridge at this location. It is a grand old structure, and when a man spends half his life working in one place, he gets attached to it. Roy's supervisors at the Illinois Department of Transportation know this. That's why he's been chosen for today's ceremony.

With the white button, Roy had set the charges; fifteen seconds later, he is to press the red button to detonate them.

". . . twelve, thirteen . . ."

As he stares at the bridge, his finger poised, a few memories of his time in the shack flash by. Most are good memories, some of them sad, some of them scary.

It's time to put this faithful old dog to sleep.

He draws a breath.

Roy Wasielewski grew up in La Salle. He's lived on the same street his whole life, all fifty-two years of it. The farthest move he ever made from home was about 100 feet away.

"I bought the house my parents owned, and after a while I bought a lot two doors down. Built a new house on it," he said. "When we moved, we didn't have to go far. It was downhill, too, so the refrigerators and stuff just rolled down to the new house."

When he was growing up, he was like most people in La Salle in that he didn't give the bridge a second thought. It was only when it stopped traffic that anyone took notice, and then they usually cursed it.

It wasn't his intent to work there. It just happened that one day there was a job opening at the bridge, and it sounded like it might be a good deal. It was certainly close to home. He reported for work in March 1975, beginning his shift at midnight. With minimal car traffic at that hour, it was a good time to start the new guy.

"That first day I had butterflies, looking at that monster of a bridge and thinking, 'Oh, my God, I gotta run that thing.'"

After two or three nights he overcame the jitters, and it became second nature to him. He liked it up there.

One of the things he liked best was the view. A person has a unique perspective from the bridge tender's shack. One sees sunsets the way others do not, and one can see rain coming before anyone else. Roy tells the story of an almost completely frozen river and two coyotes trying to cross on the ice.

"The female just ran across like it was nothing, but the male, he didn't know what to do," he says. "He kept running this way and that like he was scared. She'd kept looking back at him, like, 'Come on, you idiot.'"

Wasielewski has had some scary moments himself.

A few years ago, a towboat pushing barges of anhydrous ammonia lost power just upstream from the bridge. It was free-floating toward the bridge and coming on quick.

"Happened in the dead of night," he says. "It was pitch black."

There was nothing for him to do but sit in the shack and get ready for the jolt, crossing his fingers that the damage would be slight. Luckily for everyone, the crew of the boat was able to restore power and regain control. They passed without a scrape.

A solid hit from a runaway team of barges could have been disastrous. At the very least, the pilings could have been weakened, and this bridge might have been shut down ahead of schedule.

Before it was decommissioned, the Shippingsport Bridge was one of three lift bridges that carried vehicular traffic across the Illinois River. There are other lift bridges that support rail, but only this one and bridges at Florence and Hardin carried cars and trucks.

These bridges are manned around the clock, and they have a lot of moving parts. They are expensive to maintain and operate. Those factors and its age led the state to slate the Shippingsport Bridge for replacement, and now its wheels and motors will become spare parts for its sister bridge in Florence. A new concrete structure will cross the river here.

Wasielewski understands this reasoning. He doesn't argue with it. In some ways, it will be good for him to do something else. He's already been reassigned to another job. Change is good. At least, that's what he tells himself.

Roy exhales.

"... fifteen."

He presses the red button.

A string of orange flashes erupt like silent fireworks on the bridge. Puffs of gray smoke plume up at each

The Shippingsport
Bridge, also known as
the La Salle Highway
Drawbridge, near La
Salle. The drawbridge
was replaced with a
modern span and was
officially opened on
October 29, 2003.
The photograph was
made looking east after
heavy spring rains had
raised water levels. River
Mile Marker 224.7,
March 2001.

flash point, and then a thunderous boom cracks the morning.

The center span, still in one piece, hangs there for an instant, suspended, and then it drops like a stone-cold prizefighter, hitting the water with a splash and a whoosh.

An eerie silence follows.

The bridge is gone. Just like that, it is gone.

Someone outside cheers. Someone else blurts a laugh. Someone slaps Roy on the back. He nods and smiles. But it is awkward; the whole thing is awkward.

It was for the best, he tells himself. It had to be done. The silence returns, and it rings in his ears. And inexplicably, inescapably, he is sad.

One-Woman Operation

At the Hardin Hotel

The hotel isn't exactly on the beaten path. Tucked away on a side street up the hill in downtown Hardin, it's the kind of place one almost has to know about to find.

But that's okay for Sue Schulte, the owner-proprietor of the Hardin Hotel. She does a fair amount of business at her nine-room hotel. A few tourists find her place, usually the ones taking leisurely drives. She sees couples on motorcycles and a few cross-country bicyclists. And she'll put up a crew of dredge workers in the summer.

"It's amazing the different types of people who come here," she says.

Any more business and she'd have to hire help, because as it is, this is a one-woman operation. Schulte, who lives in the hotel, also is the front desk clerk, the groundskeeper, and the maid. She's the best employee she ever had, and the best boss, too.

The Cofferdam

A New Boat Ramp at Liverpool

Up on the levee, a cement truck idles, waiting as men with hammers set the forms at the water's edge.

From the towboat, one can see them, working with determination, a team of men, a single organism, passing boards, driving nails, dreaming of sport boats and trailers.

One of the men breaks off from the crew. He wanders to the side of the cofferdam and checks it for leaks. As long as it holds, they'll be all right. Another few minutes and they can call for the truck and begin the pour; another hour to work the cement into the forms; another day to let it set. Surely the dam will hold that long.

And the towboat powers past, pushing fifteen barges and leaving a wake. Waves and current, pressure on the dam. And the man watches it closely. Surely it will hold. Just one day. And the men work quickly, growing smaller as the towboat leaves them on the bank in its wake.

And the expectant cement truck idles up on the levee. Impatient.

*Hardin Hotel owner and
proprietor Sue Schulte,
Hardin. Parallel to
River Mile Marker 21.3,
March 2003.*

A 360-degree view from the levee on the river's west side at Liverpool. The upstream/north view is on the right side. The rear of Ruey's Riverview Inn and the public boat landing are landmarks for river travelers as they pass Liverpool's northern edge. River Mile Marker 128, July 2004.

The Stars and the Moon Above

A Commercial Fisherman's Dreams

It's after midnight on a clear night in August. The sky is full of stars, and in an hour the moon will be up. There are few things as stunning as moonrise on the river.

Jim Beasley pulls on the rope, hand over hand. Foot braced against the gunnel, he puts his back into every pull. He's fighting more than the weight of the net. There is the current, too . . . pull. This is a good one . . . pull. A hundred pounds of fish . . . pull.

And then the first hoop of the net breaks the surface of the water. There it is . . .

Beasley doesn't know what he'll haul in on any given night. It's a crapshoot every time. A commercial fisherman's nets could be fully laden, or they could be empty. There are so many variables—the strength of the current, the weather, the time of year. All of these elements play a role.

The years have taught him a few things, though, and that experience has shifted the odds in his favor. He knows that when the current is slow, his nets won't fully deploy, so he compensates. He knows that when the water is high, the fish will bunch up in the shallows.

Almost every night he reads the signs, makes a guess, and sets his nets. Two days later, he'll raise those nets and see if he guessed right. On this night, on this clear warm night in mid-August, he learns he'd guessed right.

Beasley has already dragged the first hoop of the net into the boat, then the next. His partner for the night pulls the rest of it in, and scores of fish are dumped, flopping and thumping into the belly of the boat. Buffalo, carp, and cat. It's a good haul. It's buffalo and cat he's after; the carp he throws back. The keepers are sorted and pitched into the live wells. Beasley re-baits the net and sets it again. This is a good spot.

Then he throttles up the big outboard and points the rig downriver. The twenty-two-foot johnboat skims across the water like a skate on ice. No time to waste. He has eighteen nets to raise tonight, and it will take seven hours to bring in his catch. Before dawn breaks, more than half a ton of fish will fill the boat, and there won't be enough room in the live wells.

Beasley, forty-two, runs his boat out of Grafton. From here, he has easy access to both the Illinois and the Mississippi rivers. Sometimes, when he's flush with fish, he'll sell his stock wholesale to markets in Alton or St. Louis, occasionally Chicago. But mostly he sells directly out of Beasley's Fish Market in Grafton.

One can buy fresh fish to take home at Beasley's or buy a mound of deep-fried fish modestly called a "sandwich." Restaurants buy their fish here, too.

It's a family business. His wife and high school sweetheart, Deborah, runs the market, and their sons, fourteen-year-old Jeremy and twelve-year-old John, help out wherever they can—mostly in the market

Commercial fisherman Jim Beasley prepares a paddlefish in his shop, Grafton. River Mile Marker 1, March 2002.

but sometimes with Dad in the boat. The Beasleys are a river family.

His father got him started in the early 1970s, running trotlines and selling fish to the markets. "Back then, there were three or four fish markets in town." That was in 1973. He was thirteen years old, and he's been making a living off the river, doing one thing or another, ever since. It hasn't always been fishing.

Shelling was a lucrative little business back in the 1970s and 1980s when the cultured pearl industry was paying a premium for freshwater mussels. Pieces of mussel shells are implanted in live oysters, inducing them to create pearls. Before the cultured pearl industry shifted to Asia, a few intrepid souls made a modest living by harvesting mussels in the Illinois River.

Beasley's shelling technique was crude, but it worked. He'd suit up with weighted belts, diving gear, and 100 feet of hose. A rope connected him to the boat on the surface, and as he felt his way along the bottom, he'd tow the boat along behind him. He'd come up when he had a load, and then he'd go back down. In the boat keeping watch was Deborah, his partner from the start.

"That was all B.C.," he says, "before children."

Over the years, the Beasleys became more and more reliant on fishing, and in 1992 they opened their own fish market in a little block building just off the highway in town. It wasn't the best time to go into business on the riverbank.

The flood of 1993 devastated Grafton. Whole buildings were washed away; people moved and never came back. Water filled the fish market to the ceiling, and it was closed for months. It wouldn't be the only time a flood would force the market to close, but it reopened and came back stronger each time.

Beasley is standing in the back of the boat as it skims downriver, one hand on the tiller of the 115-horsepower Mercury. At this hour—two, three o'clock in the morning—he has the river pretty much to himself.

A person can get to know the sky pretty well, working out here at this time of night, working out here mostly alone. He might not know the names of the constellations, but Beasley can gauge the time by the stars, the passing of hours and the passing of seasons.

He throttles down and glides in. He's scanning the shoreline for his marker. To remember where he's left his nets, Beasley makes mental notes of various shoreline features. It could be a power line or a cluster of buildings or the rusted hulk of an old barge. In this case, it's a dead tree on the western bank. By gauging the distance from land, he can pretty much find his nets. But this is not an exact science.

With the boat free-floating and the outboard idling, Beasley throws a grappling hook out as far as he can, and when it hits bottom, he starts pulling it back. He's fishing for the lead line of the net.

He owns about 150 nets of various sizes, with as many as 50 in the water at any given time.

A typical hoop net is a cylinder about ten to twelve feet long when fully deployed. Five evenly spaced hoops, each about four to five feet in diameter, give it its structure. The nets will vary in terms of mesh size and hoop diameter, but the basic design is the same. That hasn't changed in decades.

Attached to the hoops inside the net cylinder are mesh funnels that direct fish toward the back of the net where a bait bag is fastened—cottonseed meal for buffalo, soybean meal for catfish. Fish find it easy to swim into the funnel but virtually impossible to swim out. The net is held in place by an anchor, which is attached to a 100-foot lead line. That is what Beasley's grappling hook is hunting for and has just now snagged.

Beasley pulls the line up, hand over hand, and once he can grab the lead line, he pitches the hook into the bottom of the boat and starts dragging up the net. It's as heavy as the first one. Pretty soon the bottom of the boat will be alive with fish again, the live wells will get a little more crowded, and Beasley will be on his way to the next net.

And the moon, just a sliver of a thing, will be rising over the trees in the east.

Sometimes, on nights like this, the river will lay itself out like a flat black slate, a peaceful midnight palette that changes to indigo if one looks hard enough. It's a dark dappled blue with a streak, a speck, a blink of yellow and white, the telltale glint of the moon and stars.

But it isn't always like this. Sometimes the river can be unforgiving. It can be deadly.

"I was right there by Grafton one night, about eleven at night, me and another guy," Beasley says. "It was November."

They had a load of fish, and the wind came up. They tried to cross the river, but it seemed that everything went wrong.

"The swells were big, and when we lost the motor—right there in the middle of the river—those waves came in over the side and the boat just set up on end and went down. I lost the whole rig."

The two grabbed whatever they thought would float and hit the water. "I thought we were goners." But they somehow managed to make it to shore.

There are other hazards on the river, too—less dramatic, certainly, but this is a job where a simple injury can knock a person out and threaten his livelihood. Beasley tore up a shoulder and his back in a fall on the ice last winter. Now he has three herniated disks and no insurance. He's working through the pain for now.

"I just figured I'd be healthy till I died. I don't know, man, I see guys working jobs with paid health insurance, and they say, 'Boy, I'd sure like a job like yours, working for yourself, being your own boss.' Man, they don't know what they got."

No less a pain is the changing nature of the commercial fishing industry. People's tastes change, the market shifts, regulations get tighter, and fishing isn't nearly as productive as it once was.

"A guy like me, who has a market, we'll make it," Beasley says. "But the days of someone going out

Pauline Marchiori,
owner and operator of
Ray's Place, Hennepin.
River Mile Marker 207.5,
March 2003.

and hauling in fish to sell to the market, well, they're gone."

With the headaches involved and the future of the industry so uncertain, a guy like Beasley can have second thoughts about seeing his sons follow in his rubber bootsteps.

"I have mixed feelings about it. I mean, if my sons want to get into the fishing business, I'll help them as much as I can," he says. "But a guy has to love what he's doing to be doing this. There are no fringe benefits."

The days are long: seven hours of fighting current and getting wet. And when that's done, half a ton of fish has to be cleaned. The weeks are long: six days of hauling in nets, one day to mend the mesh. And if you get hurt, who pays the bills? And then it floods.

A person can let those worries get to him, especially while he's on land. But out here in the boat, when the air is warm and the river is calm, life's toil and troubles can just float away. He can think about his wife and his boys.

Another net is in. It's been a good night of fishing. Beasley hasn't stopped working since he put his boat in the water. But just before he kicks in the Merc and heads to his last net, he allows himself a moment to take in the sky.

See that moon? It's a sliver, smiling like the Cheshire cat in the August sky, with one bright star just off the point.

"We were out the other night, Jeremy and me," he says, still looking up at the moon. "And he said,

'I guess I've seen more shooting stars tonight than most people see in a lifetime.'"

He smiles. There are indeed some fringe benefits to this job. Tonight, in this bobbing boat in the river of time, Jim Beasley's life is a full net. And every one of the stars in the sky belongs to him.

Ray's Place

Bridging the Levee

Pauline Marchiori runs the kind of place a family might travel twenty or thirty miles to get to. Good food, friendly service; Ray's Place has that sort of reputation.

It's known by the riverboat crews, too. They know the plates always come full, and no one looks at you funny, and Friday night is the turtle special.

Ray's Place is on Front Street, right over the levee from the river landing. It bridges the lives of farm families who make their homes inland and the lives of river men who spend their days plowing water.

Fifteen Minutes in February

Hennepin Bridge in the Water

The water of the Illinois, smooth as glass, exposed for what it is: a continuum that spans a century. Trusses of tired iron are replaced by newer spans. The bridges of the last century cannot support the weight of today, and rerouted traffic passes without a glance.

The remains of the former Illinois Route 26 Bridge are photographed in a fifteen-minute time exposure at dusk, looking upstream from the east bank of the river at Hennepin. The replacement bridge, which carries Interstate 180, crosses the river and Hennepin Island. River Mile Marker 207.6, February 2001.

Beyond the light of day, with shutter wide, a camera can capture what the eye can't see. A window into darkness. Night becomes day, and time blurs.

In fifteen minutes, one hundred years can be exposed, and these trusses are a bridge to yesterday.

An Evening in Marseilles

The Kat Nip Club and a Thirst

He's a thin man with sharp angles. As he cranes his neck to draw another drink from the bottle, one can see it go down, can almost taste it for him. It's a cold one at the end of the workweek. It's good. Yeah, life is good.

His name is Jim, and he's a machinist for a small company just north of town. He's a hardworking man who's seen some ups and downs in his life. One can tell just by looking at him. He's not more than forty years old, but his face is deeply lined with stories. It's dark in the tavern, but a person can read a lot in a face like that.

So what's there to do in this town on a Friday night?

"You're doing it," he says. He ought to know; he's lived here most of his life, and he's a regular at the Kat Nip Club. "Wednesday night is dollar beer night, and Thursdays are fifty-cent drafts."

A bar in a town like this does what it can to draw customers. It's fairly empty for a Friday night, but so is the town. A bar has to rely on the locals, on guys like Jim, to stay in business.

A lot of faces like this can be seen in taverns like this, in towns like this all along this stretch of the river. There are stories of hard times and good times etched into those faces, stories of opportunities that left town, stories that raise hopes, too. Surely something better will come along. Any day now.

Marseilles could be any number of river towns that grew to prominence because of the proximity of the waterway, towns that declined when the currents of commerce shifted. The boats got bigger, but they brought less business to town. The railroads quit taking on passengers, and the interstate—the latest nail in the coffin—sucked the economic life out of downtown.

A lot of these towns have turned to tourism. Some are better at it than others.

There's an effort to cash in on the designated National Heritage Trail of the Illinois & Michigan Canal. Lockport, Channahon, Morris, Ottawa—they're all paying attention to the old canal. In Marseilles, the canal bed is weedy and as dry as a parched throat at the end of the workweek.

Jim's bottle is empty. If a person buys Jim a beer, he'll buy him one back. It's a code, just what is done. It keeps everything even. But not everything in life is as cleanly defined or as fair.

As he puts it, Jim's got six kids, three of them his own. He's got a girlfriend and an ex-wife he's still in love with. He's got a little house and big responsibilities. It can make a man old before his time.

He's had his chances to move on and might have done better for himself if he had. He's seen one job or

another wiped out from plant closings and such, but he chose to stick it out here. "This is my home," he says. No better reason.

Marseilles, too, has had its own opportunities over the years. It was the first town on the Illinois River to harness the waterway's generating power. In the 1830s, a dam was built to divert water for a sawmill operation. Later, a hydroelectric plant was built along a diversion channel—one of two in town—and it generated power for neighboring towns and the old interurban train line that operated along this stretch of the river. That plant still stands, but it's fallen somewhat into disrepair, and weeds get tall and thick around it in the summer.

In recent years, industry has bypassed Marseilles. The high school fell victim to declining enrollment and closed in 1992. Some of its students went to Ottawa; some went to Seneca, a smaller town.

The largest employer is Field Container, a cardboard box manufacturer. Before Field, it was Nabisco. And it was Federal before that, or maybe after. One loses track around here. With each change

A 360-degree time exposure in downtown Marseilles. The center of the photograph, just over the shadow of the camera and tripod, is looking west. River Mile Marker 247, July 2004.

of ownership, there was a disrupted workforce. It's hard to keep a good-paying job in town.

Some of the former employees in Marseilles decided to go to work for a plant in Morris, another paper-product manufacturer. That's where they developed the self-locking box. At one time, the Morris plant made all the Chicken McNugget boxes in the world. There are some in Marseilles who envy Morris that distinction.

During the day, the locals down at Betty's Cafe will state exactly what's wrong with the town and

its elected officials. "They let the school close down," says one.

"They won't fix the streets, except if they're up there where the politicians live," says another. "The downtown's gone to hell."

But in softer moments, perhaps after a second cup of coffee, they'll reveal why they've spent their entire lives in town and would never think about moving. They say they love it.

The evening is over. It's almost midnight, time to head on home. So Jim says goodnight and walks out into the street. It's Friday night, and it's quiet in Marseilles.

A town will show a different face at this hour, just as people do. And one can read a lot in it. There's an odd joy in the air. Contentment, one might say. And there's a regret or two in the wind, as well.

Silent Passing of Steel

Replacing Bridges

Methodically, the bridges built in the 1920s and 1930s are being replaced. Hennepin, Shippingsport, Morris, they've all come down in successive years. Those bridges were built at a time when the Illinois Waterway was starting to take shape. Locks and dams were going in, and it was time to build. Now it's time to replace.

In Morris, one two-lane span of the new bridge was constructed before the old steel structure was removed. Another two-lane span will complete the replacement.

The demolition was a delicate operation. Workers had to protect the new bridge, sometimes using sheets of plywood to shield it near the detonation points. And they had to quickly clear the old bridge from the navigational channel. There were no hitches, no delays. It went smoothly, which is to be expected—the crews have had some practice at this sort of thing lately.

The Morris Highway bridges, looking southwest, on March 11, 2003, the day the main channel span of the old bridge was demolished. The Illinois Route 47 changeover to the new span had occurred earlier but was temporarily halted during this cold spring morning for safety's sake. The front structure, finished in 1934, can be seen intact in the earlier photographs in the sequence. The targeted span had been partially dismantled and the decking was removed prior to the systematic explosions that would bring down the span. For a view north over the site after the explosion, see page 185. River Mile Marker 263.5, March 2003.

Part Two

Take any road you please . . . it curves always, which is a continual promise,

whereas straight roads reveal everything at a glance and kill interest.

—SAMUEL CLEMENS, "Some Rambling Notes of an Idle Excursion"

Looking northwest into downtown Chicago as the deck of the Columbus Drive Bridge, with a horizontal span of 177 feet, rises during the annual spring trial openings. The Michigan Avenue Bridge, in its lowered position, is the next bridge downstream. Chicago River Mile Marker 326, April 2001.

A Road Journal

The Michigan Avenue Bridge is a good place to start.

There is a view from here. And history.

Plaques embedded in the bridge towers commemorate a host of events, from Marquette and Jolliet's arrival to the reversal of the Chicago River. Fort Dearborn, which protected the early settlement until it was attacked, once stood near here; a plaque in the sidewalk says so. The towers of this bridge, built in 1920, are adorned with bas-relief sculptures that depict the early struggles of this frontier town—at least from the white man's perspective. And now, a new museum in the tower has opened.

It's a well-traveled bridge. Thousands of people drive and walk across it every day, and most of them don't give a second thought to the significance of this place. Perhaps they don't care. Perhaps they've simply forgotten.

I pause to read the plaque dedicated to "those pioneers who plied this water route." It reads: "This river . . . linked the waters of the Atlantic, the St. Lawrence and the Great Lakes with those of the Illinois, the Mississippi and the Gulf of Mexico. From 1673, commerce and civilization . . ."

A man's voice interrupts.

"Excuse me, sir, excuse me," he says. It's rapid-fire; he's selling something. ". . . but I've been looking at your shoes. See how bad they look?"

Well, yes, they could use a little attention. The guy who has just pointed that out, of course, is carrying a box of shoe-shining equipment. Now, what did that plaque say about commerce . . . ?

"Let me show you what I can do with this one shoe." So I prop a boot up onto his box. "It won't cost you a dime."

"Yeah, but how much will it cost me to get the second shoe shined?"

"Five dollars and a tip. The tip is up to you."

. . . and what did it say about civilization?

So I become another customer of Sam the Shoe Doctor, a young man who is as much a fixture at the Michigan Avenue Bridge this summer as one of its plaques. He slaps on the polish and buffs the leather with a small soft brush, finishing his work quickly. It's a decent job.

I pay him five dollars and tip him a buck, and he goes off in search of another pair of dusty shoes. I return to the plaque.

"*. . . commerce and civilization followed this natural waterway from the seaboard to the heart of the continent.*"

It's hard to imagine what this river was like back then, before all the concrete and steel, back beyond the façade of history. It's easy to lose the thread that binds us to those early settlers, the explorers, the first inhabitants of the marshes and river banks.

It's easy to lose sight of the fact that the water that rolls under this bridge will roll under sailboats, pleasure craft, canoes, barges, and bridges all the way to the Mississippi River. It's easy to forget that these waters link us across time and space.

With those thoughts, I turn to leave. But I don't get far.

"Excuse me, sir." It's Sam again. "Excuse me, but I've been looking at your shoes. See how bad they look?"

He's already down on his knee and pulling out brushes. "Now let me show you what I can do with this one shoe. It won't cost you a dime."

"Sam, Sam, we've already been through this. You just shined my shoes."

He looks up. "How long ago was that?"

"Probably a minute or two."

How easily we forget. He smiles sheepishly and slowly picks himself up.

"Sorry about that," he says. Right away, though, he spots another pair of possibly scuffed shoes. "Excuse me, sir, excuse me . . ." and off he goes.

People can experience the Illinois Waterway many ways. They can float it, fly over it, walk its banks, bike its trails, and drive across it. Each way provides a unique perspective.

From the air, the valley is a strip of greens and blues, varying in width and density. Silent, tiny boats and the wakes they leave can be seen. One can get the big picture from up there. But it is a cold, sterile view, and none of the other senses is engaged.

On foot or bicycle, people can get a closer look from the towpaths and banks. They can smell the water and vegetation and watch the water flow in slow motion. The experience is immediate; it's in their face. But it is limited to one small section of river. It leaves a person hungry for more.

By boat, people can feel the river breathe, can hear it lapping, splashing, giving way, and pushing. They can anticipate the landscape before it unfolds, then watch it slip away, and it's gone before they are ready for it to go. It is an intimate experience. But the vision people have on the river is focused fore and aft, channeled by the banks and levees, and they never see the connection between river and town, between water and bottomland. They don't see the creeks, the swamps, and the backwaters that keep this river alive. They don't meet anyone.

In a car, sometimes drivers zip by with the air conditioner blowing and the radio blaring, and they don't even notice the river unless the bridge is out, or up, and they can't get across. This is how most people see the river—which isn't seeing it at all.

But there is another way to drive this river. It's along the designated byways and down the gravel roads to the boat ramps and dusty dead ends. And if travelers take that route, from Chicago to Grafton—and if they take their time—they can get a sense of the big picture as well as of the smaller ones.

So with the road to discovery lying ahead, I turn my taillights to the Great Lake and point myself downstream toward Grafton and the Mississippi River. I'm headed down the Illinois Waterway—the Chicago River, the Calumet–Sag Channel and the Chicago Sanitary and Ship Canal, the Des Plaines and finally the Illinois River. I'm headed to the heart of the continent, and I really don't care if my shoes are shined.

FROM THE CITY

Keeping the Chicago River in sight while driving out of the city can be a trick. For one thing, the traffic demands attention. If you wanted to spend a few hours making ninety-degree turns every two blocks or so, the Chicago River can be crossed about twenty times before one gets out of the Loop. It might take awhile, but it would give a person a good look at the waterfront renaissance downtown.

Once the river turns southwest, it melds into the straight and businesslike Chicago Sanitary and Ship Canal. Interstate 55—the Stevenson Expressway—parallels the waterway here, riding directly on top of the abandoned Illinois & Michigan Canal. This is industrial territory: lots of rail yards and loading docks, power stations and wharfs.

Alongside the Stevenson on the south is Archer Avenue, which goes through old ethnic neighborhoods and villages—from Bridgeport, where the I&M Canal had its start, to Summit, which was the hump those early canal builders had to get over.

Across the Stevenson from Summit, the Des Plaines River drops in from the north. Right over there is where Marquette and Jolliet portaged into history a few hundred years ago. Except for small boats, the Des Plaines isn't navigable here. It turns west and skirts the working canal for about twenty-four miles, to just below the Lockport Lock and Dam. That's where the Des Plaines becomes deep enough to handle the shipping chores.

But this stretch near Summit, the old I&M route, is like a conduit carrying all the utilities to and from the city's southwest side. Along this narrow corridor run the Sanitary and Ship Canal, the interstate and a state highway, power lines, pipelines, and miles and miles of rail—the Burlington Northern Santa Fe and the Illinois Central. It's an urban umbilical cord.

Once Archer Avenue leaves Summit, it cuts through Willow Springs and into a vast green expanse, a forest preserve of bluffs and ponds and picnic groves. After the clumps of tank farms, container trucks, and rail yards, this comes as somewhat of a surprise. The Palos Forest Preserve is sliced in straight lines and angles that define the boundaries of subdivisions and canals. Its overall shape is triangular with its apex formed by the meeting of the Chicago Sanitary and Ship Canal on the north and its Illinois Waterway brother—the Cal–Sag Channel—on the south.

Turn Basin No.1 on the Calumet River near
95th Street. Calumet–Sag Channel Mile
Marker 332.3, March 2000.

"Here, hold this."

Jerry Dykstra hands me a rod and reel while he picks
up another and sets the hook. Before he can reel in his
fish, I get a strike, and a couple of minutes later he has
two more perch in his bucket and both lines back in
the water.

"Yeah, they're thick in here," he says.

Dykstra, a thin, weathered working man from
Lansing, Illinois, looks to be in his fifties. For years
he's been fishing off this bank near South Ewing at the
mouth of Calumet. It's a place called the Tenth Ward.

Politically speaking, the Tenth Ward is made up of
a large piece of the south side of Chicago. In the par-
lance of fishermen, though, the Tenth Ward refers to
this particular spot, this chunk of concrete, within cast-
ing distance of the bridge's shadow and within sight of
the lake. When he left home this evening, Dykstra told
his family he was going to the Tenth Ward; they knew
exactly where he'd be.

When the perch are biting, it's difficult to find room
to cast here. It's a popular spot, and the fishermen line
up shoulder to shoulder. But this is midweek, and word
of the perch run hasn't gotten out yet. Dykstra reels in
another fish. It's a bitty one, barely six inches long. He
tosses it back with a farewell word. "Go on, get bigger,
come back and see me."

He's got a bucketful of keepers already, and he's going to have to clean them all.

"The kids are coming up," he says, referring to his grown children and their appetites. Have to put food on the table this evening.

Does he ever wonder where this river leads? Does he ever get the urge to take a boat downstream to see where it goes?

"Oh, sure," he says. "Sometimes we put in at 95th and go down to 106th Street."

Curiosity has its limits. But why go any farther if you can fill a bucket right here?

The Calumet River is not glamorous. From very early on, it was a working waterway. In 1870, it was identified as having more potential as a port than the Chicago River because its mouth wasn't plagued by sandbars, which tended to clog the Chicago, and because it was broader and discharged more water.

So a deep river channel was dredged, and the port took shape. The bulk of the lake shipping shifted there, away from downtown. As the city grew and the steel mills fired up, this place saw a lot of activity. Ore was lugged down from the northern Great Lakes and off-loaded for the mills. The port is still one of the busiest on the Great Lakes, and a lot of grain and scrap is shipped out of here. Steel and liquid bulk, in various forms, come and go.

But the mills are cold, and the iron ore boats from Superior don't line up anymore. The land is being reclaimed, and a park is taking form where the old U.S.

Steel plant once stood on the lakefront just north of the mouth of the Calumet. Gray-brown fields to greenways; times have changed.

The area around the Calumet Harbor is where George M. Pullman built his empire of railroad cars. You can still see the remnants of the "perfect town" he built in the late nineteenth century. The neighborhood of Pullman, once an independent municipality, was a planned company town that stood as a shining beacon of capitalism for about two decades before it was destroyed by a depression, a strike, greed, and other economic factors.

The Calumet, like the Chicago, was reversed to flow in from the lake. It runs south, doglegs right, then left, and opens into the industrial harbor. On the banks, heavy equipment works around huge piles of material—scrap iron, gravel, and sand, the raw products of progress, or perhaps survival.

This is where the large Calumet sewage treatment plant is, where wastewater is treated and released into the waterway. And just across the expressway from the plant is the Thomas J. O'Brien Lock, the first lock on the Cal–Sag arm of the waterway. It's the controlling works, regulating how much lake water is diverted downstream. At night, when the lights are trained on this structure, it is an imposing sight, an all-night operation.

Unless one is on the Calumet Expressway or a thoroughfare like Torrence Avenue, driving through this area is like trying to navigate a concrete labyrinth full of potholes and pipelines, railroad tracks and manufacturing plants. This is a dirty brick place with

overpasses, viaducts, and manholes. The streets make ninety-degree turns, and some of them just end.

But tucked into this protective industrial armpit, shielded like some sort of urban Shangri-la, is the blue-collar community of Hegewisch. It's a place of well-tended lawns and small, clean middle-class homes. The streets are straight and tree-lined, and there's a faint, fading Eastern European feel to this neighborhood.

Dean Ubich sets a cold beer in front of me and wipes the bar one more time. It's a compulsive gesture; the bar is already clean. It is polished and smooth, a testament to his work ethic and to good towels.

The South Shore Inn on Brainard Street is a quiet little tavern that probably hasn't changed a lick in forty years. It's virtually empty tonight. With its nondescript signage and gentle ambience, it feels more like a barbershop than a pub.

"We get a lot of regulars," he says. "We're just a neighborhood bar."

Ubich has the manner of a benevolent brother-in-law. His smile comes easily, and he's not in a hurry to pick up the dollar I've set in front of me. He leans against the back bar and swaps a story or two.

Heading upstream in the Chicago Sanitary and Ship Canal, under the Ninth Street Bridge in Lockport, deckhands begin to break the barge load so that they may drop off barges before making the "turn" at nearby Lemont. River Mile Marker 292.5, November 2000.

The South Shore Inn is a family business. His grandfather opened the place just after Prohibition, and it's been in the family ever since. Ubich took it over in the 1970s, after college, and he and his family live in the house next door. It's the house he grew up in.

Lining the walls of the tavern are vintage posters of the old South Shore Line, the railroad that hooked around the southern tip of Lake Michigan, through Hegewisch, bringing tourists and workers to the city in the early 1900s. The bar's sleek curves and frosted glass give the place an art deco flavor, resonant echoes of a Century of Progress. A felt pennant from the 1959 White Sox is tacked up behind the bar.

"I feel like I'm operating a museum sometimes," he jokes.

As far as bartending, Ubich keeps to the basics. A person can get a shot or a beer here, maybe a simple mixed drink, but don't go asking for one of those fancy concoctions with a cocktail umbrella.

"We just don't do that," he shrugs. "Oh, I might have some Grenadine here somewhere."

When business gets slow, Ubich just locks the front door and turns off the lights. He tells me this, and I look around. Then I start feeling like I'm keeping him up.

So I finish my beer, wish him well, and head out the door.

"Thanks for stopping," he says. And he still hasn't picked up my dollar.

Beyond the O'Brien lock, the Calumet River runs into the Little Calumet, which straightens out and becomes the Cal–Sag Channel, which runs dead-on to its juncture with the Chicago Sanitary and Ship Canal.

The Cal–Sag is so named because it links the Calumet River system with an area known as the Sag Bridge, which lies within that triangular Palos Forest Preserve. Along the way, the waterway slices through a mixture of industrial and tightly packed residential areas—Blue Island, Crestwood, Alsip, Palos Heights, Palos Hills, and Worth, communities along Illinois Route 83, known locally as the Calumet Sag Road.

There are a few green areas here, small forest preserves and the groomed grounds around the SEPA plants. Slowly, the landscape is shifting, becoming less suburban and more townlike. And once you reach Sag Bridge, you've left Chicagoland.

TO THE MEETING OF THE WATERS

Outside the grasp of the city, the highway becomes the Joliet Road and passes through the historic towns of Lemont and Lockport, where remnants of the I&M Canal are well preserved—as they should be. After all, these towns owe their existence to that canal, the first manifestation of Louis Jolliet's vision and the predecessor of the Illinois Waterway.

Lemont is a clean little town that is using its historical tender to barter with tourists. The old canal still holds water. There's a pleasant little park, quaint buildings, and nice restaurants downtown. And churches—there seems to be a church of every color and stripe here, one on every other corner.

There's some muscle to this town, too, activity down by the working waterway. As a terminus for barge traffic, Lemont is prime ground for industry. Northbound boats make the turn here, handing off clusters of barges to smaller boats for the narrower passage into Chicago.

Along the waterway south of town, tank farms and oil refineries—Unical and Citgo—are clustered. Pipelines and conveyor belts cross overhead while tandem trucks and semi-tractor trailers push product on the roadway and rail cars line up on the siding.

The road to Romeoville intersects about halfway to Lockport. Taken to the west, it will quickly carry you over the old canal, the waterway, and the Des Plaines River, all running parallel here. At the river lies the Isle a la Cache Museum, which offers an informative overview of that time when Native Americans first interacted with the Europeans, the French voyageurs.

Romeoville is home to Lewis University, which keeps a voluminous archive of I&M Canal documents and research materials. Its map collection has no peer. Just south of the university is the road back across the waterway and into Lockport.

Lockport, as the name suggests, was key real estate in the canal era. The town was laid out by I&M commissioners, who located their administrative headquarters here in 1837.

A well-preserved section of canal runs just below State Street and the canal museum. Down at the canal itself, the impressive stone Gaylord Building stands, as it has for more than 150 years. Now it is a visitors' center, a restaurant, and the Illinois State

The Cass Street Bridge is one of five bascule bridges in Joliet. Upstream on the Des Plaines River, a barge load is being pushed through the open Jackson Street Bridge, five city blocks north. Des Plaines River Mile Marker 288.1, March 2002.

Museum gallery. The town, with a generous collection of Victorian homes and well-maintained downtown structures, is kind of a gallery in and of itself.

Lockport has the flavor of a larger Lemont and serves as a transition to the much larger Joliet, which lies next door. These three municipalities, besides being products of the I&M, are bound by something else: stone.

More than fifty quarries were operating between Lemont and Joliet while the canals were being built. The limestone was used to construct the locks and canal walls, and many of the buildings in these three towns are made of the same material. The Gaylord Building is one. The stone wasn't the finest building material ever pulled from the earth, but for decades it was used in factories, houses, hotels, schools, and prisons that still stand after a century and a half. It was a prized material as Chicago rebuilt after the fire in 1871.

One of the most imposing stone structures is at the edge of Joliet, the old Joliet Correctional Center. The newer Stateville Prison, on the other side of the river, is made of the same stone.

Joliet is a big city compared to Lockport and Lemont. Downtown buildings are proud, romantic edifices. A seawall contains the waterway, and a nice, clean park overlooks the river, which is crossed by a set of beautiful green bascule bridges downtown. In the heart of downtown is Harrah's Casino and Hotel. A second casino, the Empress, is just down the road.

Although the canal came through here, Joliet owes as much to the railroad industry for its development.

Before Chicago, Joliet was the region's rail hub. In the 1850s, the Chicago and Rock Island and then the Chicago and Alton Railroad serviced this city. In the 1890s, the Elgin, Joliet and Eastern line set up headquarters here. Nearby coal fields spurred the steel industry, which fed the railroad boom, and the economy kicked into gear.

Tough times hit Joliet in the twentieth century, and vacant homes line the wide and hollow streets on the bluff north of the river—large turreted houses that once demanded servants to maintain them. Joliet has rebounded somewhat in recent years, but the economy seems to be powered more by the city's prison industry and its gambling boats.

The I&M crossed the Des Plaines River at Joliet. Since its origins in Chicago, the canal had followed the river on the south and east side. But it jumped across to the north in Joliet and stayed there until it reached La Salle, its terminus.

A traveler can cross, too, and stick close to the river, riding U.S. Route 6 into Channahon.

Anyone interested in canal history would enjoy walking in the park here. The old lock walls have been maintained, along with a lock tender's house, one of two that survive in the Illinois & Michigan Canal National Heritage Corridor. The I&M towpath is a well-tended trail here, following the banks of the ever-widening Des Plaines. Seasoned hikers and bikers can follow this trail all the way to La Salle.

Channahon, which took its name from the Potowatomi word for "meeting of the waters," lies at the confluence of the Du Page and the Des Plaines

The Elgin, Joliet and
Eastern Railroad Bridge,
looking directly north
across the open lift-span
to the tracks on the
north side of the river.
The main channel span
is typically seen in the
upright position as the
line is only occasionally
used for freight traffic. (See
also page 43.) River Mile
Marker 270.6, January
2001.

rivers. The name might also refer to the river that comes in on the far side just downstream, the Kankakee. Where the Kankakee and the Des Plaines meet, they officially end. And the Illinois River formally begins.

MORRIS TO MARSEILLES

"That bridge, the one they're replacing, was the second steel bridge over the river," he says.

Ken Sereno is in his living room in Morris spilling dates and tidbits of yesteryear onto the table. He doesn't have to look anything up. He has absorbed it all because for the past few years, he has been studying local history like a man obsessed.

"I was in business for forty-eight years," he tells me. "Started a blacksmith shop on my twenty-first birthday. That was in 1947. But then I hurt my back and got into the Ace Hardware business."

He worked long days and long weeks. Work was all he knew. He had no hobbies. When retirement came, he wasn't ready. He didn't know what to do with himself. "Truth is, I worked for free for the fella I sold out to. The thought of going home scared me. I needed something to do."

So he started going to the library to chase away the demons of boredom.

"The first bridge was made out of wood . . ."

And the more he learned about Morris's history, the more he wanted to know.

". . . before that, there was a ferry going across there from 1841 to 1856. It was run by the Armstrongs . . ."

He scanned microfilm of old newspapers—the Morris Herald, *the* Gazette—*some going back to the 1850s. He spent four years at those microfilm machines. He recites names and facts in a dry, matter-of-fact tone, as if reading a term paper.*

". . . the first steel bridge used the same supports the wood bridge used. It went up in 1900 . . ."

His newfound passion led him to write a book, 150 Years of Progress: Grundy County Grows with the I&M Canal. *Throughout the community, he is known as the town historian.*

". . . nothing wrong with that bridge; it was just flat. So when they raised the level of the river, they had to replace it. That was in 1933."

Sereno also can tell you about Chief Shabbona, the Potowatomi leader who warned the white settlers about an imminent attack during the Black Hawk Wars in the 1830s. "Saved a lot of lives."

He can tell you more about the Armstrongs, who donated land for a new courthouse while retaining ownership of all the adjacent downtown property. "They weren't dumb."

He can tell you about the canal, the railroads, the first corn festival in 1949, the oldest homes in town, the long-closed Gebhard brewery, and just about anything else older than last week's newspaper.

He can tell you all of that and still shrug about being called "town historian."

"No," he says, "I just needed something to do."

Although all the towns from Lemont to La Salle trace their roots to the I&M Canal, often it was agriculture

that sustained them. Lumber from cleared farmland and grain from the fields came into town to be
shipped out on the canal. Elevators popped up, and
the towns grew. Here in Morris, the Illinois River valley turns to farming.

The road from Joliet is lined by new housing subdivisions and townhouse developments, and after
Channahon it passes under an umbrella of power
lines streaming from the Dresden Nuclear Power
Plant south of the river, but when you enter Morris,
the whole landscape changes. This is a large mid-
western rural community, not unlike any found in
central Illinois or Iowa.

Three large grain elevators dominate the riverfront,
and because a lot of corn and beans move through
here, farm-related businesses thrive. The downtown
square, home to the Grundy County Courthouse, has
a prairie town ambience, and the townspeople have
that slow, easy manner that invites you to sit a spell
and have another cup of coffee. That's the way it is,
anyway, at Weit's Café, the "Home of Good Food."

Morris has a lot going for it: proximity to Interstate
80, major secondary highways, a new river bridge,
a beautiful section of the I&M Canal, a state park,
marinas, hiking trails, and some light industry. It's a
diverse economy, something not all towns along this
stretch can claim.

During construction of the canal, towns sprang up
along its route. Some—Morris, Seneca, Marseilles,
Ottawa, Utica, and La Salle—thrived at various times
and to varying degrees. Others are merely ghosts
today and footnotes in history. The towns of Des

Plaines, Dresden, and the original Kankakee (the
first town of that name was at the confluence of the
Kankakee and Des Plaines rivers) did not survive.

The high ground and bluffs above the towns give
way to rich farmland to the north. It's fertile soil here,
and the agricultural influence noticeable in Morris is
apparent all the way to Spring Valley.

U.S. Route 6 can carry you efficiently all the way
there, cutting across vast, verdant fields, past large
farms and clusters of blue Harvestore silos. But there
are more adventurous routes, ones that sweep closer
to the river. The Old Stage Road, which runs from
Morris to Seneca, is one. Stagecoaches that linked
Chicago and La Salle used this trail in the early nineteenth century. The stage was the best mode of travel
during the canal's construction and before the railroads came through with a vengeance. The road hugs
the river and the canal on the way west.

Seneca is a town of about 2,100 people and "Home
of the Fighting Irish," or so it says in a mural on the
wall of the Blades & Bullets bait shop. It is a tidy
little community with an unremarkable and quiet
downtown—which is a far cry from sixty years ago.
During World War II, the shipyard down by the river
employed more than 11,000 people as the boat makers
worked around the clock to churn out troop-carrying
landing craft for the military. There isn't much left of
that shipyard today.

Across the old steel bridge, though, there is a lot
of activity. A group of marinas, one after another,
services yachts and pleasure craft of all sizes. There's
Hidden Cove Marina, the Seneca Yacht Club,

The main channel portion of the former Morris Bridge lies in the river, photographed from the (temporarily intact) south deck, looking north toward Morris minutes after the bridge was demolished. (See pages 168–69.) The new bridge was protected during the explosions by four-by-eight sheets of plywood, a necessity evidenced by the large impact scars on the upper sheet in the foreground. River Mile Marker 263.5, March 2003.

A valve used to manipulate the level and flow in the diversion channel next to an old manufacturing plant in Marseilles. River Mile Marker 247, March 2003.

Custom Marine, and Spring Brook Marina. The places are well-tended and the grounds manicured. Seneca's proximity to Chicago and the availability of dependable and affordable "sales and service" attract a lot of boaters. Plus, this is a reliable river pool, extending from Joliet to the dam in Marseilles, the next town downstream.

Marseilles was spotted here because the long series of rapids made this a logical stopping place for early river travelers. A dam was built in 1830 to provide hydropower. The concept was expanded, and diversion dams were installed to send water through two manmade channels that looped through town. One of those channels directly powered industry, most notably the Nabisco plant. The other ran to a hydroelectricity plant, which supplied power to the old interurban train that linked the towns in the valley. Both of those channels, as well as the I&M Canal in Marseilles, are dry today.

When the modern waterway was being built in the 1930s, the old dam was replaced and a lock installed. Because the rapids are so long and Marseilles lies so close to the water, a conventional lock and dam would have been impractical, so a new navigation channel was dug beyond the south bank, in effect creating a thin three-mile-long island. The dam blocked the old channel and kept the town dry, while the lock was in the new channel at the downstream end of the island, beyond the rapids.

South of that channel and stretching along the bank is Illini State Park. You can park an RV or pitch a tent here and listen as the riverboats push through the lock. The next morning, you can drive west on the leisurely and winding Gentleman Road into South Ottawa.

OTTAWA TO STARVED ROCK

John Nordstrom has something he wants to show me outside. Besides, activity inside the tent is starting to pick up and it will be easier to talk out there.

The tent is revival-size, large enough to host a small circus. A morning breeze sweeps through the open walls, ruffling the banners behind the altar, and overhead the broad expanse of canvas ripples as if stirred from slumber.

A dozen volunteers and church elders are deploying folding chairs, setting up flowers, serving doughnuts. Service will begin in about half an hour. "Where is that box of bulletins?" someone asks. "Does anybody know?"

A young volunteer peers under a table. "Nope."

The Christian rock band Audio Adrenaline goes through a sound check. "Testing . . ." and a guitar player strums a raucous chord.

It's 8:30 in the morning.

Nordstrom is the leader of this congregation, Christ Community Church of Ottawa, which has been meeting all summer under the big top on the south bank of the Illinois River. The building the congregation had called home is being sold, and for reasons involving federal funds and the separation of church and state, the church had to vacate. Their exodus from the building could be a temporary situation, but nothing is

From the west side of the Illinois Route 23 Bridge, on the north side of the river in Ottowa, this 360-degree image was made looking south toward Allen Park. Known formally as the Veterans Memorial Bridge, it connects Columbus Street in Ottawa to State Street across the river in South Ottawa. The Fox River, a significant tributary, can be viewed joining the Illinois River from the extreme left side of the photograph. River Mile Marker 239.6, July 2004.

certain. A few people's faith is being tested this summer.

But a lot of things have just fallen into place—the availability of this land, for one. And other congregations in Ottawa have been supportive. And on the day it became clear the congregation would lose access to its building, a visitor from New York state just happened to know where to find a large tent.

"We are making the most of it," Nordstrom says.

He has tailored some of his sermons to tap the allegorical power of the river, drawing analogies to the flight from Egypt and coming to the river Jordan. They've had an occasional baptism in the river, too.

"We'd just finished setting up the tent—this was Memorial Day weekend—and one of the members saw a turtle come across the grass. It came from the river and was crossing right in front of the tent. And it stopped right here."

Nordstrom points to a tent stake. This is what he wanted to show me. A thick, taut rope runs from the stake to the tent canvas, and a short plastic fence encircles it, creating a protective zone about four feet in diameter.

"Amid all that commotion and all the work we were doing, right in the middle of all this confusion, she chose to lay her eggs right here."

It was a sign that they have come to the right place, he says. It gave them hope that life would prevail, despite the hardships. And that turtle provided him a sermon or two.

Ottawa has as strong a link to the I&M Canal as any town, including Lockport and Chicago. Ottawa and

Chicago were the first two towns to be platted by canal commissioners, who intended the two towns to be the anchors of that waterway.

At that time, 1830, Ottawa was to be the terminus of the canal, probably because the Fox River empties here into the Illinois, which became wider and deeper as a result. But still lying downstream were the rapids at Starved Rock, and clearer heads eventually prevailed, so it was decided to extend the canal to La Salle. Not much remains of the old canal here, which passed just north of downtown. But there are reminders.

The street named Canal, for instance, runs along the route of the old dogleg that jutted perpendicularly from the canal and carried goods to and from downtown. And the aqueduct that carried the canal over the picturesque Fox has been restored. It is one of the canal's most striking surviving structures.

Ottawa has a big city feel, with a downtown full of stately buildings, some five stories tall. The La Salle County Courthouse of cut stone is an imposing presence. There are many large restored brick structures, and the stone here is of a finer quality than what is in Joliet.

Besides the canal, a lot of history swirls around this town. It was home to W. D. Boyce, who founded the Boy Scouts. And two years before the Civil War, Abraham Lincoln and Stephen Douglas held one of their famous debates here. About five miles west of town is the site where, in 1673, Marquette and Jolliet paddled into a huge settlement of Illinois Indians. Friendships were forged here, and tribe members would later lead the Frenchmen upstream and show them the portage into the Chicago River.

The site of that Indian village is off of Dee Bennett Road, which starts in Naplate and heads west to Utica, passing old silica mines, Buffalo Rock State Park, and the Starved Rock Lock and Dam. The Illinois Waterway Visitors Center, operated by the U.S. Army Corps of Engineers, is attached to the lock and dam and houses exhibits about the waterway. Replicas of a towboat's wheelhouse and Native American artifacts share space in this facility, which also offers a wide-angle view of the modern waterway's lock and dam in action.

Dee Bennett Road is a pretty drive, but if one is looking for natural beauty, the road of choice is on the southern bank, Illinois Route 71. It follows the sweep of the river and then winds through Starved Rock State Park, a crown jewel of the state park system.

The park is full of nature trails, hidden canyons, and bluffs. There is a lodge built by the Civilian Conservation Corps in the 1930s, campgrounds, and spectacular scenery. It also comes with a legend.

The rock itself, a large outcropping at the edge of the river, was the scene of a grisly battle between warring tribes of Native Americans. As one version of the legend goes, a band of about 1,200 Illinois Indians, blamed for the death of Pontiac, the Ottawa Indian chief, took refuge atop the rock and were surrounded by Ottawa, Potowatomi, and other allied Indians who sought revenge. The Illinois Indians were slowly starved out. Some tried to fight; others tried to slip away. None survived.

It's a good story, but one that probably is more myth than history.

UTICA TO LA SALLE–PERU

A smiling Luis Lugo leans back on the bench, his arms folded across his considerable chest. He's on the verge of laughter, clearly tickled by the story he's telling.

"We'd go anywhere to fish," he's saying. "We once drove six, seven, eight hours to get somewhere in the middle of the night, and there wasn't a motel. We'd be tired and feeling down, and we'd be, 'Oh, man, that's bad; oh, man . . . oh, well, let's go fishing.'"

And he lets loose a belly laugh, one of those deep, infectious, reverberating laughs that you can almost see rolling all the way down the riverfront.

Lugo and a fishing buddy drove from Chicago to Starved Rock State Park this morning. It's only a couple of hours from home, so they won't need a motel room tonight. They've made this trip many times.

"I just love to fish. And every Friday, that's fishing day," he says. Lugo is a minister, pastor of Monroe Street Church of Christ on the west side of Chicago. "I tell my congregation that if they have an emergency, don't have it on Friday."

Lugo has come many times to this bench by the park's visitors center. There is a long walkway with a railing and several benches at river's edge. You can see a good section of the river in both directions. The lock and dam is off to the east on the opposite bank. Lugo's pole is propped up against the railing, a line feeding down to the water. He's hoping the stripers are biting.

"What I like about this spot is you meet some of the nicest people. They come from every facet of life, every nationality. And then you see people whose lives depend on fishing." He nods in the direction of a couple about twenty yards away. "I fish because it relaxes me. But those people are fishing because they have to."

The two—a thirty-something man and a slightly younger woman—look hungry. They are leaning over the rail and checking the deployment of their lines. They have five lines in the water, two five-gallon buckets at their side, and a look of determination—not quite desperation—on their faces.

Lugo, a native of the Virgin Islands, knows that look. As a youth, he fished every day in the ocean, sometimes all day, he says. If he didn't catch anything, the family went hungry. The laughing minister is suddenly quiet. There is a depth to the silence.

Prodded, he tells how as a boy he came to the United States with his family, how he grew up in the city, and how somewhere along the way he found his calling and became a minister. So, I ask, does he consider himself a fisher of men, too? He shrugs.

"I don't know about that," he says. "All I know is fishing is good for you. As a matter of fact, I tell my congregation that very thing. I tell them, 'If you don't fish, it's a sin. There's no way you are going to heaven.'"

And there's that laugh again.

From the park, a 1930s-era steel bridge carries you across the river and up the bluff to Utica, population 1,000. Technically speaking, this town called Utica isn't really Utica at all; it's North Utica. The original Utica, which long ago succumbed to flood and disinterest, was down by the river; it no longer exists. When the canal made it this far west, the people of old Utica moved up the bluff to meet it.

Canal construction uncovered rich veins of mineral deposits all along its length, and this area was particularly rich. Quarries and mines were quickly opened, pulling silica, limestone, coal, and clay out of the ground. Some of that material went into constructing locks and buildings; some was shipped out on the canal at Utica, Ottawa, La Salle, and Peru.

Downtown Utica is one street and a few storefronts. There are a couple of taverns, a restaurant, and the La Salle County Historical Museum, which is housed in a restored I&M building. The community was devastated in 2004 by a tornado that took eight lives, and the effects of that storm are still evident downtown.

The Canal Road, running along the old towpath, leads west to La Salle. It would be a mistake to think of La Salle and Peru as one town, although many people do. They share a high school, after all, and an invisible municipal boundary. But there is a strong civic pride in each town, which is one of the reasons why each has a river bridge.

The differences between the two are many. At the core, though, the basic distinction is simple: La Salle is a canal town, and Peru is a river town. What was Canal Street in La Salle is Water Street in Peru.

View from the river looking north. The lights of Peru (left), La Salle (right), and passing towboats can be seen. The location is just upstream from the Illinois Route 251/Joseph M. Wallin Memorial Bridge, near the river bend that rounds the Huse Lake area. A long exposure resulted in a blurred photograph as the boat rocked due to fleeting traffic. River Mile Marker 223.8, October 2000.

La Salle is the bigger of the two towns and, like Ottawa, feels urban. Large brick and limestone buildings anchor the downtown, which commands the high ground, well above the Illinois River. First Street has a row of nice shops, and old stone storefronts line Marquette Street. The old Rock Island Line railroad station is now a lawn and garden center. The terminus of the canal—at one time either the starting point or the destination for thousands—is now a park.

Peru, by contrast, is more active down at the river. Railroad tracks still trace the bank. Towboats can tie up, and barges can load. Mertel Gravel Company is down here, along with Maze Lumber, which has been in business since 1848, as its sign says, doing "what ever it takes." Peru, clearly, is a town that works for a living. And like so many towns downstream, it works at river's edge.

COMING 'ROUND THE BEND

One of the river's most distinctive physical characteristics is the Big Bend, the sharp southerly turn caused by glacial encroachments and thaws of a different age. But it marks a shift in more than the direction of the current.

The valley becomes woollier here. It's less industrial, less agricultural. It is untended. That's the way it is on both sides of the river all the way to Peoria. There are pockets of shallow backwater lakes, some of them vast expanses that open off the river channel—Goose Lake, Swan Lake, Senachwine and Sawmill lakes. And there are woodlands, too. This is part of the Mississippi Flyway, the route taken by hundreds of thousands of migratory birds each year. This is hunting territory.

Private clubs and public waterfowl hunting areas dot the shorelines. The state operates a public area at Depue Lake and the fragmented Marshall State Fish and Wildlife Area. The towns of Hennepin, Henry, Lacon, and Chillicothe are vested in the river and these natural areas.

U.S. Route 6 leads straight west out of Peru and into Spring Valley, a pleasant town well off the river and up the bluff. Main Street is wide and lined with two-story brick buildings. Spring Valley has a gentle, small-town ambience, and one half-expects to see Harold Hill and his band of trombones come marching down the block at any minute. Indeed, the Municipal Band plays downtown every Friday night in the summer, closing each show with the "Star-Spangled Banner."

There's a river bridge here, a 1930s steel truss that has the distinction of being the last bridge that crosses the Illinois on a north-south orientation. But a bridge alone doesn't make a community a river town. Spring Valley is more of a coal mining and railroad town.

Illinois Route 29 begins here, and it sweeps around the top of the Big Bend, past the tiny towns of Depue and Bureau, skirting the Lake Depue State Fish and Wildlife Area and across the Hennepin Canal, the short-lived and ineffectual link between the Illinois and the Mississippi rivers.

The landscape is decidedly scratchier here, like a day-old beard. The farmland is uneven, irregular in

places. The vegetation isn't mowed with the same frequency as the land to the east; it's wilder. And now, you have a choice.

If you were to stay on Route 29—it's called the Ronald Reagan Trail through here—you would drop straight down past backwater lakes, through Henry and Chillicothe and into Peoria, following the flow of the river on its western bank. Or, you can cross the river, pick up Illinois Route 26 at Hennepin, and travel the eastern bank through hunting areas, down a narrow green corridor into the town of Lacon, and then on toward East Peoria. Both are

pleasant drives in the evening with the sunset on the river, which is ever widening through here. Or, you can do a little of both, crisscrossing the river on the bridges at Hennepin, Henry, and Lacon.

ON THE EAST BANK

The road to Hennepin used to cross a steel bridge right into the town, but Route 26 traffic was funneled onto the new Interstate 180 Bridge just to the north, and the old bridge was removed in 2001.

The main channel span of the former Illinois Route 26 Bridge at Hennepin was taken down on March 22, 2001, about an hour before this photograph was made. Immediately after the bridge was brought down, the work to extract the segments from the river began. The U.S. Coast Guard allowed a brief closure of the channel. This view is to the south from the Interstate 180 Bridge. The village of Hennepin can be seen on the left. River Mile Marker 207.6, March 2001.

Hennepin is one of the oldest communities on the Illinois River, founded in 1817, a year before Illinois was granted statehood. The town was named for Father Louis Hennepin, who was with the well-financed French explorer Robert Cavalier Sieur de La Salle as they expanded on the explorations of Marquette and Jolliet in the late seventeenth century.

Today it is the prototypical river town, a must-stop for barge operators. Large grain elevators line the riverfront, and there is a boat launch and a store that caters to river workers. There are towboats docked here and barges tied off across the river. This is a small farm town on the landward side, but it has the feel of a busy port at the waterfront.

The next town on this side of the river, Lacon, is a busy river town, too. Since its founding in 1831, it has been a grain-shipping center, and there always are barges tied off to the shore waiting to receive from the ADM Growmark facility. As the county seat, a lot of business is transacted here. It supports a country club, the Marshall County Historical Society, and tidy homes for both young families and retirees.

The road south toward East Peoria is a pretty drive of about twenty-five miles, occasionally taking you close to the river and through a series of wildlife areas. Just before the Peoria area metropolis is the town of Spring Bay.

Little more than a gaggle of homes and small businesses today, Spring Bay was the first town settled in what is now Woodford County. It was touted as an important port at one time, and some contend it might have rivaled Peoria if all the railroad tracks had not been laid on the other side of the river.

ON THE WEST BANK

If, instead of crossing into Hennepin, you stayed on Route 29 on the western bank, you would land in Henry, founded in 1834 and the site of the first lock and dam on the Illinois River. Remnants of the old structure can be seen down near a modern marina. The lock and dam helped make this town because all river traffic was funneled through here and forced to stop. When the lock was removed for the new waterway, though, the town slipped.

This is also home to Steamboat Elsie, the legendary riverboat lover who petitioned the state back in the 1930s to retain the lock. She didn't win that argument, but, undeterred, she shone a light and hailed boats with a megaphone as they passed, a practice that lasted years. Her home still stands, and so does the little stone lighthouse she had built as a tribute to the river men.

Henry has a tight two-lane river bridge, which had to be closed for a short time one recent June when its lights attracted thousands of swarming mayflies after their annual hatch. When the short-lived mayflies died, the highway surface became dangerously slick.

The road south from Henry runs past a refuge and a fish and wildlife area and into Sparland, population 504, where a bridge crosses to Lacon. South of Sparland is Chillicothe.

Chillicothe, "Chilli" to the locals, is a long, skinny town that hugs the river on a stretch of high ground. The downtown is several blocks long with a combination of small businesses, pubs, and old storefronts, many of them vacant.

Like so many towns through here, it was platted in the mid-1830s and shipped out farm goods on the river. It boomed when the railroads came through in the 1840s and 1850s. The Rock Island Line, which ran passengers and freight between Peoria and Chicago, was first, but it was soon joined by others, including the Atchison, Topeka and Santa Fe, which came in from the west. The Santa Fe built a drawbridge across the Illinois River in 1888. That bridge, which was replaced in 1931 by a 440-foot-long steel truss bridge, lent the city its motto, "Where the rails cross the river."

It is at Chillicothe that the river balloons into Peoria Lake. Route 29 streams past Rome—another former rail town and now a slender bedroom community of lakeside cottages—and Mossville. By the time you arrive in Peoria, you've had a good, long look at the wide expanse of the lake and the river channel marked off by buoys. It is a deceptively shallow lake.

THE SHORES OF PEORIA

Jim Baldwin is standing above the river on an old concrete spill. It resembles a gray lava flow with tufts of grass poking up here and there. It is otherworldly. This is just north of the riverfront, close to the Murray Baker Bridge. He looks out at the water.

"*The river is something Peorians have taken for granted for so long," he tells me. "They don't recognize how important it is. This area was settled because of the river, and this city was built around these lakes."*

Baldwin, a retired vice president of Caterpillar, has been watching the shoreline and Illinois River for decades. He can vividly recall when the riverfront was a mess of rail and industrial waste, a dumping ground. The concrete flow he's standing on is an example of that.

But he's also seen a remarkable turnaround down here.

When he was still working for Caterpillar—for decades the heart that pumped this city's economy—he served as chairman of the Peoria Riverfront Business District Commission, a coalition of government, industry, and citizen concerns that was charged with revitalizing the city's waterfront.

Slowly but surely—with a blend of government grants, corporate donations, and the hard work of individuals—the riverfront was resuscitated. Today it is a healthy vital organ of the metropolis. Restaurants, nightclubs, meeting halls, art galleries, shops, a museum, and a dock for an excursion boat—the Spirit of Peoria—*command almost a mile of shoreline.*

The view southeast from North River Beach Road in Rome on the morning of a crisp New Year's Day. River Mile Marker 177.5, January 2003.

Looking into Peoria and downstream from the middle of the Interstate 74 Murray Baker Bridge the day after the Peoria portion of highway and this bridge were closed for reconstruction. River Mile Marker 162.3, April 2005.

"The idea was to bring people down to the river again." Apparently it's working.

Financing for the redevelopment wasn't unlimited, though, and sometimes the planners had to get creative. A hiking path connects the riverfront to a new fitness center, winding along landscaped lots that used to collect industrial debris. Most of that junk has been removed, but the concrete flow remains.

"We thought about what to do with this, but it was going to be too expensive to remove," he says. "So we just incorporated it into the landscape. We call it the moonscape. Children like it."

Baldwin also is the executive director of the Heartland Water Resources Council of Central Illinois, a private organization focused on the health of Peoria Lakes. He has seen troubling signs of sedimentation in the lakes, but he finds hope in the riverfront.

"It's only been in the past few years," he says, "but people are beginning to see the importance the river still has to Peoria."

Peoria is called River City. It is the city of bridges, the city Cat built. It is a rail hub, an educational center, Whiskey Town, the oldest settlement in Illinois. Discounting Chicago, Peoria is the largest city on the waterway. More than 100,000 live within the city limits, and it is part of a metropolis three times that size.

Peoria—named for the Peoria tribe of Native Americans who lived along the river here for hundreds of years—was the first permanent European/American settlement in Illinois. The French explorer La Salle built a fort near here in the 1670s. In the early nineteenth century, when steamboats began churning up the river, it developed quickly as a port. Later it became a rail center, with fifteen separate railroads operating in the city.

A lot of coal and grain was shipped out of here, too, and it was that synergy—a relationship of energy, farm, rail, and river interests—that spun off other industries and ignited growth. Cheap and abundant grain, for instance, made Peoria a prime location for stockyards. So many distilleries, also big consumers of grain, sprouted up that the city was soon known as the whiskey capital of the world.

Heavy equipment manufacturers, most notably the behemoth Caterpillar, are headquartered here, but the economy has shifted toward the service and medical industries. The medical centers of Methodist, St. Francis, and Proctor are collectively the city's largest employers today.

A drive through Peoria shows a little of both worlds. Newer, gleaming buildings dress the central business district. There are hospitals and clinics and a convention center downtown. On the south side, down at the river, lie the older manufacturing plants. These are tall and dark brick factories with smokestacks, some of them cold and silent.

The city is alive with activity. A minor league baseball team plays out of a new sports complex. There's a minor league hockey team, too. Bradley University is here, along with the University of Illinois Medical School. There's a federal courthouse, and this is the county seat of Peoria County.

Two boys enjoying an early spring day on the Pekin Riverfront Pier beneath the John T. McNaughton Bridge. Upstream of the bridge is Cooper's Island, home of the Pekin Boat Club. Even farther north, on the west side of the river, are the stacks from the Edwards Power Station. River Mile Marker 153, March 2004.

This is a city of bridges, too. The McClugage Bridge, the first crossing for vehicles since Lacon and Sparland, carries U.S. Route 24 across the narrows between the Upper and Lower Peoria lakes. Interstate 74 crosses the river on the Murray Baker Bridge, and there are three more bridges leading out of the city.

It's impossible to isolate Peoria from the rest of the metropolitan area, though, which includes Peoria Heights, West Peoria, and Bartonville. Directly across the river is East Peoria, home of a riverboat gambling boat, a big Cat facility, and a riverfront park of its own. There are also the cities of Creve Coeur and Pekin. The entire area is home to about 350,000 people, and, civic pride notwithstanding, it is a single throbbing metropolis.

But when you leave this place—either by U.S. 24 on the west or by the back roads on the eastern bank—it doesn't take long for the urban to give way to the rustic. You have entered yet another distinct region of the Illinois River valley. And the area you are entering might have no counterpart anywhere.

On the western bank, U.S. 24 steers out of Peoria, heads through Bartonville, and sidles up to a row of gentle bluffs. Large modern industrial plants squat between the highway and the river, but as the town of Kingston Mines zips by in a blink, all of that industry is left behind. Soon you are cruising into Banner Marsh State Fish and Wildlife Area.

This is a large tract of wetlands and prairie, open for hunting and fishing. Directly south of Banner is another wildlife area, Rice Lake. These natural areas prime you for the Nature Conservancy's Emiquon preserve, the 7,000-acre farmland-turned-ecology project, about twenty-five miles down the road.

On the road toward Emiquon, row crops fill the bottomland to the left. The corn comes in waves, flowing into the valley from above the bluffs. Acres and acres of corn and beans now, green on green, rolling into troughs and swelling over hillocks, an ocean on the banks of the river.

Liverpool, population 150, is an old settlement, a dead-end town at the end of a three-mile road. In the beginning, its people lived off the river. They hunted, fished commercially, and harvested mussels. But that way of life changed after the backwater lakes were drained, particularly Thompson Lake just to the south. Surface water was diminished and sedimentation increased, destroying spawning ground for fish. Aquatic vegetation, which provided feed for migratory birds, died off. Consequently, waterfowl and fish pop-

ulations plummeted. This is a common theme on the lower Illinois. The effects are seen first in Liverpool.

The most vibrant business in Liverpool is Ruey's Riverview Inn, a nice restaurant and bar that sits up against the levee in a corner of town. It has been here a long while, operating under different names—Ruey's now, and Helle's before that—but always called the Riverview. Lining the walls are photographs of old steamboats and hunters and fishermen showing off their kills.

Before the levee was built, you could actually view the river while dining at the Riverview. It's a little touch of irony that punctuates the fact that although the levee protects the town from flooding, it is part of the backwater drainage project that killed the town's hunting and fishing culture.

Back on U.S. 24, the highway slips through the community of Little America, which is gone in the time it takes to snap off a salute. Picking up Illinois Route 78, you leave U.S. 24 and head south. To the right is the road to Dickson Mounds State Museum and National Historic Site. Straight ahead is the bridge that crosses to Havana.

While traveling from Peoria to Havana, it's not easy to keep the river in view from the roads on the west side, but it's even more difficult on the east side. A series of county blacktops zigzags through the back country. You skirt Sand Ridge State Forest and end up going through two little towns named Goofy Ridge and Buzzville, which make Liverpool feel frenetic by comparison.

This road runs alongside Lake Chautauqua, part of the Chautauqua National Wildlife Refuge. This is a haven for migrating waterfowl, a much-needed resting spot for the birds. There are white pelicans and ducks of all sorts, blue heron and geese. There are songbirds and eagles. This is a large body of water.

Travelers need a map, but they can eventually reach Havana this way.

Steve Moehring is in Waterworth's Restaurant and Lounge telling a story on himself. It's about the time he was out with his wife in his new boat.

Moehring, a banker at Havana National Bank, acknowledges that he didn't fish much or hunt and had never owned a boat. But for some reason, at this stage of his life, he felt he needed one.

"Everyone I knew had a boat. That's just what you're supposed to do in Havana, right? So I went out and got one."

It was a used johnboat, an eighteen-footer with a 50-horsepower outboard motor. Well, one evening he and his wife, Kay, took it out. They stayed on the water later than they probably should have, but it was a pleasant night. The air was cool, the river calm, and they were just free-floating near the far bank close to the bridge. Ah, peace. What could be better than this? This must be why people own boats, he thought.

It was right about then when the unmistakable sound of a towboat came from around the bend. That was the signal to end the evening.

"We figured we'd better get out of there because he might not see us, and barges can't stop anyway, so I try

to start the motor," he tells me. "Nothing happened. I try it again. And again. It was dead."

Meanwhile, the barges had rounded the bend. Moehring, his heart pounding, grabbed the paddle and thought briefly about trying to cross. Better not. So they tucked the boat against the far bank and waited for the barges to pass.

Once the wake had quit rocking them, they tried to cross the river. It wasn't easy. They had to battle the current in order to not overshoot the dock.

"I was paddling so hard, it was like one of those cartoons," he says, laughing about it now. "First on one side, then the other."

They finally made it, but the experience left him with a new appreciation for the river—the current can work with you or against you. And it left him with a new respect for barges and towboats—they can be relentless and cold. And it made him worry about what went wrong with the motor.

"Turns out there's a loose wire," Moehring says. "It was just a silly little thing."

Havana is the capital of waterfowl on the Illinois River. This is home to the Illinois Natural History Survey, the Forbes Biological Station, the Chautauqua National Wildlife Refuge, Emiquon, the Nature Conservancy, federal fish and wildlife offices, and the Illinois Department of Natural Resources.

During the town's annual Illinois River Festival, the "Quack & Honk Meat Calling Contest" is held. This is a non-sanctioned event. The Havana High School's athletic teams? The Ducks.

Although the town has a lot of life today, with a balance of new and old residential areas, a busy downtown, and a diverse culture, it was a veritable beehive 100 years ago. When commercial fishing and hunting enterprises propelled the economy from 1900 to its peak in the 1930s and 1940s, Havana was on the verge of becoming a full-throttle, full-blown city. At one time, there were four floating fish markets at the riverfront and five grocery stores in town. There were three movie theaters, three drugstores, and an assortment of gambling houses, hotels, and bars.

Today on the river, there's a private marina just across from where the Spoon River flows in from the opposite side, and there are public boat launches and several large grain elevators downtown. Just south of downtown is the Ameren Illinois Power plant. This is a coal-burning facility with huge black stockpiles reaching to the roadway. No matter where you are on the river, it seems, you are not far from either a power plant or a grain elevator.

INTO THE WILD

There are two roads south out of Havana, and each offers a view of one of the wildest areas of the valley.

On the east side, Illinois Route 78 takes you quickly through Matanzas Beach and follows the contour of the river to Bath before dropping cross-country to Chandlerville on the Sangamon River. There you pick up the Beardstown Road and take it west, back to the Illinois.

On the west side, you take Illinois Route 100, dubbed the Illinois River Road. It follows the western bluffs, which are becoming more pronounced, past Anderson Lake and the small towns of Bluff City, Sheldons Grove, Browning, and Frederick before delivering you across the bridge and into Beardstown.

It doesn't matter which route you take; the landscape is wild and the people are self-reliant. All the way to Meredosia and beyond, the sloughs, chutes, and backwater lakes have bred generations of hunters and independent souls.

There is a direct link, in many cases familial, to those days when people lived off the river and the wild land along its banks. Trapping, hunting, and fishing are still primary activities here, but it is rare that anyone can make a living off the river today.

In Browning in the early twentieth century, for instance, commercial fishermen accounted for one-third of its population, which at that time numbered about 500 people. Today, the River's Edge Tavern, which sits on stilts about twenty feet in the air to keep it dry during floods, is full of hunters and fishermen. But it's strictly sport and food for the table these days, although every now and then someone or other has been known to go over the limit.

This is the territory of the old market hunters and the unrepentant poachers, and that spirit still thrives in these backwaters. Little lakes and swamps dot the lowlands here. There are areas where a person can slip away and be willfully lost for days. This is the land where Al Capone and other gangsters retreated to when it got too hot for them in Chicago. Locals

Looking north and upstream at the Frederick Landing Boat Launch in Schuyler County. During many months of the year, pickups and trailers will be found at the landings for boaters along the lower portion of the Illinois River. The locations to "put in" along the northern stretches—for example, the Des Plaines and the Chicago Sanitary and Ship Canal—are not nearly as plentiful or as easily accessible as many areas in the rural downstate sections. River Mile Marker 91.6, September 2004.

still talk about those days, and they talk about them proudly.

There are little clusters of houses and cabins, some of them year-round residences, back in those areas. Patterson Bay and Snicarte near the Sanganois State Wildlife Area are two such communities. These are backwoods hunting towns.

THE LINCOLN TRAIL

This area is the heart of the region settled during the great Kentucky migration in the 1820s and 1830s. From Liverpool and Meredosia, backwoods-savvy

Kentuckians slammed together towns and went to work on and along the river. Abraham Lincoln, whose name crops up with greater frequency on this portion of the river, was part of that migration.

After a stint in Indiana, Lincoln's family settled in the Sangamon River valley not far from here in 1830. Two years later, he was piloting boats down the Sangamon to the Illinois. Although the Sangamon proved to be unnavigable for larger vessels, a couple of steamboats tried to make it work for a while. Lincoln worked on one of them, the *Talisman*, in 1832.

There is a monument in Bath commemorating Lincoln's "House Divided" speech, which he delivered here in 1858. That was the same year of the celebrated "Almanac Trial" down in Beardstown, where Lincoln discredited a witness by using a farmers' almanac to refute his testimony. The crime that precipitated that trial—a fatal beating inflicted in a drunken brawl that broke out after a revival—was committed in Bath. There is a museum where the trial took place in Beardstown, and there are still a lot of taverns in Bath.

Bath sits on a chute that used to be the main channel of the river. Floods long ago gouged a deeper channel farther to the west, creating Grand Island and leaving the town sitting on the narrower chute. Like a lot of towns here, this place was once jumping, with steamboats docking and commercial fishermen pulling out tremendous catches. Today an exclusive private hunting club is located on Grand Island.

South of here, the river veers away from the highway to the west. About fifteen miles down the road, past the cutoff to the Sanganois State Wildlife Area, you cross the Sangamon River and ride into Chandlerville. Here, you pick up the Beardstown Road—designated the Lincoln Heritage Trail—and follow the Sangamon back to the Illinois and Beardstown.

The Sangamon is channelized and is basically a ditch through here. Despite that, it's still a pretty approach into Beardstown. The broad valley is quartered off into neat acreage for crops, and for miles the road winds along the tree-covered bluffs and postcard-pretty farmhouses that were built 100 years ago. Eventually the road reaches Beardstown and guides you onto East Fourth Street.

CHANGES IN BEARDSTOWN

A wooden walkway along the flood wall in Beardstown provides an ample view of the river. It is one of the few places you can actually see it in town. With the flood wall sealing off the downtown, it's possible to forget the river is even there.

Harold Worley is here at the Beardstown River Look almost every day. He likes to see what's happening on the river—today there's some dredge work being done down at the bridge. With him is his dog, Pablo, who is straining to be somewhere else.

"Stop that, Pablo," he scolds, tugging a little on the leash.

Worley has a mental disability that prevents him from working a normal job. He has difficulty figuring out basic tasks and following directions, but he enjoys photography and has a pretty good memory for certain things. He's friendly and talkative. And he'll tell a stranger—like me—what he can about the old bridges here.

"Used to be a wooden bridge that crossed right here," he says. "I can't tell you for sure when they torn it down, guy, but the relics of the old bridge used to be across the river right over there."

He points to a spot on the far bank.

"Don't know if they're still there. . . . Will you knock it off, Pablo?"

*The William H. Dieterich Memorial Bridge
at Beardstown, looking southwest during the
winter of 2000. This winter was extreme, and
the river closed to commercial boat traffic
for several weeks. The temperature was six
degrees on this day. River Mile Marker 87.8,
December 2000.*

Talk turns to his hometown. Does he like it here?

"Oh, yeah," he says. "It's a nice place to live. I mean they put this park in here, and this walkway. This is really nice." We agree on that.

Pablo is tugging at the leash, though, and it's time to head down the walkway to get a closer look at the dredging. He bids me a friendly good-bye and slips away.

The flood wall has a slogan painted on it at this location. "River of Dreams," it reads. But the decking of new walkway cuts off the bottom of the letters, creating the impression that the sentiment is sinking.

Beardstown is a town in transition, but perhaps it has always been that way.

It grew and suffered as a lot of river towns. It was founded around a ferry crossing in the 1820s and quickly blossomed into a thriving river town. It harvested the river: commercial fishing, shelling, and ice cutting. Blocks of ice were cut from the river, packed in sawdust, and stored in icehouses. In summer, the blocks were sent out by barge or rail. Ice-cutting was a major industry for many towns on the river, from Peru in the north to Beardstown in the south, until 1909, when artificial ice-makers became common.

One industry that has always been influential is meat-packing. In the mid-nineteenth century, thousands of hogs each year were processed at the Beardstown slaughterhouse and shipped out on steamboat. Today, the town's major employer is the Excel pork packing plant. The packing industry, though, has little to do with the river these days because most of its product is trucked out.

Excel has brought the latest huge change to town. Since moving into a plant vacated by Oscar Mayer in 1987, Excel has recruited cheaper nonunion laborers from Mexico. Consequently, the town's Hispanic population has soared. In 1990, there were 56 Hispanic residents in town; ten years later, there were more than 1,100. That's almost 20 percent of the town's total population.

Signs in Spanish adorn business fronts, schools have adjusted curricula, social service agencies have responded, and the Catholic Church is suddenly cramped for space on Sundays. The demographic shift has had a few rough spots, but overall Beardstown seems to have adjusted. And it has moved another step away from its reliance on the river.

The flood wall, built in 1928, provides the metaphor.

A marina north of town and Logsdon Tug Service keep the town's connection to the river alive, as do the grain elevators. But the riverboats don't dock downtown anymore, and except for the view at the River Look, one can't even see the river from downtown. A voice on the educational tape at the Beardstown Museum says, "The high flood wall . . . is an almost symbolic separation of the town and river that gave it life."

THE ALTERNATE ROUTES

From Beardstown, there is any number of routes to Grafton.

Illinois Route 100 will get you there in one piece, for instance, but not necessarily in a straight line. It crosses the river twice on the way and takes several inland detours. It is a relatively easy route and leads through some picturesque country and the towns of Pearl, Kampsville, and Hardin. But that route, the designated Illinois River Road, steers away from the water in places, bypassing the important river town of Meredosia and ignoring some quirky places like Valley City and Montezuma and Brussels and a host of natural sites on both banks.

You could choose to stay strictly on one side of the river, or you could just make it up as you go along.

On the western bank south of Beardstown, a back road crosses the twisting LaMoine River and winds along the foothills past Spunky Bottoms, the 1,157-acre restoration project of the Nature Conservancy. Along the way, you pass the La Grange Lock and Dam, the last (or first, depending on your outlook) lock and dam on the Illinois River.

Illinois Route 99 crosses the river to Meredosia, but if you pick your way south along the western bluffs, for the next fifteen miles the view of the ridges to the right and the lush bottomland to the left is splendid, and these roads eventually lead to the Ray Norbut State Fish and Wildlife Area. This is a gem of an ecological area tucked under the twin Valley City Eagle Bridges, which carry Interstate 72 overhead.

These roads also lead through Valley City.

Donna Westfall and Kay Taylor are planning a party. It should be one to remember.

Westfall and Taylor are co-owners of Sleepy Jack's in Valley City, and the party they will throw in three weeks will be their last. The tavern is closing.

Sleepy Jack's is a dimly lit little pub in an old storefront in a shell of a town in the middle of nowhere. It has a loyal clientele of locals, lost salesmen, and bikers.

The name of the pub is conspicuously painted on the front the building, but it's hard not to be conspicuous in a town with only half a dozen buildings. A year ago it had one more building, but fire destroyed it. It was a tavern, too.

The blaze eliminated the women's only competition, but ironically it signaled the beginning of the end for Sleepy Jack's. Shortly after the fire, someone organized an effort to declare the town dry. The women hired a lawyer and fought back.

The fight had the look and feel of a family feud, an intra-family feud, because everyone in the town seems to be related to each other. The tavern that was burned was owned by members of the Westfall family, Donna's in-laws by marriage. There are also several Westfalls on the village board. Kay used to be married to a Westfall, too, and a former brother-in-law to both women, another Westfall, is the third partner in Sleepy Jack's.

"It is kind of complicated," Kay acknowledges.

There were twenty-three eligible voters in the election, eighteen of whom voted, ten of whom voted for a dry town. The women contested the results. Then they resigned themselves to defeat. The doors will close in three weeks, and that will be that.

I ask Donna what she'll do after she shuts down.

"I guess I'll go on welfare, just like everyone else around here."

But before that, there's a party to plan. And here in Valley City—on the back road to nowhere, in a town of twenty-three eligible voters—it should be a party to remember.

PRESERVING HISTORY

The road south from Beardstown on the east bank follows the bluffs all the way to Grafton and the Mississippi River. The vast farm fields of Scott and Greene counties stretch out between the road and the river, which is held back by levees the entire way.

Along this route are cutoffs to the towns of Meredosia, Naples, and Florence. Meredosia is about twelve miles south of Beardstown. It has a steel truss river bridge built in 1936, a power plant, a grain elevator, and the Meredosia river museum.

The museum is run by Dora Dawson, a passionate historian and lover of the Illinois River. She will gladly walk visitors around the exhibits and explain how the old-timers used to harvest the river. Some of those old-timers are still around and sometimes demonstrate their folk crafts in the museum. Sometimes they find a willing student and take them on as apprentices in a program funded by the state.

Want to learn how to weave a hoop net or build a basket trap? It can be learned here. You can also see how they used to harvest the mussel beds and hunt mink, beaver, and muskrat.

"Pretty soon, all of the old-timers will be gone," Dawson says. "We're just trying to preserve the memory a while longer and maybe pass the knowledge on to another generation."

Jeff Barnett is sitting inside the River Museum, a trail of twine and knitted nets around his feet. He's weaving a hoop net the old-fashioned way. He pauses long enough to do the math.

More than fifty dollars for the hoops, twenty dollars to have them delivered . . . about twenty-five dollars for three pounds of twine . . .

And the time?

About twelve or fourteen hours to knit the net . . . another three or four hours to put it all together . . .

And how much to buy a new net from a commercial fishing supplier?

"It ain't worth it," he says. "But this is a lost art."

Barnett learned this art from Earl Edlin, a longtime river man and self-named "river rat" who was at this very moment in another room of the museum teaching the skill to a teenager from Jacksonville, part of another generation of hoop net makers.

Barnett grew up on the river, in Beardstown, and although he makes a living as a lineman for the Ameren power company, he carries a commercial fishing license and boasts of a long family connection to the river life. "My grandfather made a living on this river," he tells me, "just like Earl." He adds, "I don't particularly care to go to an art museum and look at all the pictures. I'm not into all of that weird art that some people like to go and buy. This is my kind of art."

A small boat travels west from the Illinois River into an inlet that has been constructed around the pipes that run from Pike County's McGee Creek Drainage and Levee District, just southwest of Meredosia. River Mile Marker 71.2, September 2004.

A truck moves west across the Florence Bridge. The lift bridge is operated by the Illinois Department of Transportation. An operator is stationed in the control house twenty-four hours per day. The vertical clearance above pool stage is 26.6 feet when down and 83.4 feet when the bridge is raised. The photograph was made on the upstream side of the bridge from one of the protective concrete cells placed on the east side of the lift-span. River Mile Marker 56, March 2001.

Meredosia was a terminus for Illinois's first railroad, the Northern Cross, which began running a twelve-mile route in 1838. It was touted as the first railroad operating west of the Allegheny Mountains. The line quickly extended to Jacksonville and Springfield, and eventually it was linked to the state's expanding rail network. A depot built in the 1880s still stands in town.

Naples is just south of Meredosia. It is home to a boat ramp, a couple of grain elevators, and a collection of homes today, but in the first half of the nineteenth century, the town was an important port, as was Florence, another town that took its name from an Italian city. Florence, Illinois, bears no resemblance to the Italian city of art, but it can claim possession of at least one treasure: the river bridge, which is one of the two remaining highway lift bridges on the river. The other is in Hardin.

The Florence Bridge today carries Illinois routes 100 and 106, but for the longest time, this was the U.S. Route 36 Bridge, known informally as the Coast to Coast Bridge because it replaced a ferry and became the last section of unbroken pavement of a road used by some to travel cross-country. U.S. 36 now shares a crossing—the Valley City Eagle Bridges—with Interstate 72 north of here.

The town is down on the water, with houses strung along like beads on the bluff. Down toward the southern edge of town, Troy Hawkins is bringing up a turtle from the river bank. It must be two feet long and heavy. Hawkins, who runs the River Rat Den bait shop, sees dinner in this turtle. This one will be soup tonight.

Illinois Route 100 leads to Pearl, but it leaves the river for a while. The county roads, though, stay closer to the water and thread through Montezuma and Bedford and some of the prettiest country in the valley.

If you pull off the road and rests awhile on a small hill overlooking the bottoms and the river, near Buckhorn Creek, you can hear a chorus of summer insects, the call of birds, and the drone and growl of a towboat churning up the channel, just beyond the trees.

THE DAYS OF FLOOD

Back in the old shelling days, the harvested mussels would sometimes yield a gem. The town of Pearl took its name from that rare and wonderful discovery. This town, like so many others, rose on the hopes that lay in the river.

The button factory in Pearl spurred the town's growth. Meredosia, Beardstown, Naples, and Kampsville had similar factories. The remains of Pearl's old factory are out on the highway at the edge of town. Close by is Pearl's newest business, Big River Fish, a processing plant that could be the future for this town.

On the road again, the bluffs close in on the river. These bluffs are part of the ridge that runs like a spine down the length of Calhoun County, which begins just south of Pearl. The county lies entirely between the Illinois and Mississippi rivers, a thin peninsula whose creeks drain both watersheds.

There isn't a single railroad in the county. It wasn't worth the cost of laying track to push the railroad down that peninsula, and the people didn't need rail because they had easy access to the rivers to ship out their fruit. This is prime orchard country.

Calhoun is bordered on the north by Pike County, and the only way into Calhoun is along that twenty-mile border, or across the bridge at Hardin, or by ferry. Two ferries cross the Illinois River—at Kampsville and at the very tip of the county, near Brussels. Both run every day, around the clock. The state operates them, and they are free.

Kampsville was formally settled in the 1860s, sprouting up around a private ferry and a store. It prospered during the steamboat days, profited from the shelling craze, and got a boost after one of the first locks and dams was located here. That dam and the button factory are long gone, but the old Kamp Store is still here. It is on the National Register of Historic Places and houses the Center for American Archeology and a museum, which exhibits artifacts of the valley's earliest inhabitants.

About 200 people live in town, and the largest employer is Louie's Kampsville Inn, down at the ferry landing. Two dozen people work at the restaurant and bar. It's open seven days a week, unless it floods. When it does, the business can be closed for months.

The Illinois Department of Transportation ferry at Kampsville is one of two ferries on the Illinois; the second is twenty-nine river miles downstream near Brussels. The ferry at Kampsville connects Illinois Route 108 from the east to either Illinois Route 100 (north/south) or Route 96 (west, out of Kampsville). This view is to the east from the Kampsville landing during the high waters of 2002. The rising river's edge required a temporary extension of the walkway, shown being built on the left. River Mile Marker 32, June 2002.

The town's growth has always been tempered by flooding. There is no levee here. Louis Hazelwood has owned the Kampsville Inn for eighteen years, and he's seen six floods—four in the 1990s, another in 2003. It can wear on a person.

"I'm closed for months. You have to replace carpeting, tile, everything," he says. "And when you reopen, it takes a long time for people to discover that you're open for business again. I'm a Pisces, and I like water, just not all of it."

Bertha McGowen sits in the shade of her fruit stand just south of Kampsville. It's late June, dry, warm. She fans herself and returns to her tale.

"If you talk to anyone in Calhoun," she tells me, "the way they distinguish time is before the flood and after the flood. Life was different after that."

McGowen, sixty-one, is recalling the 1993 flood, the big one, the one that wiped out East Hardin across the river and flooded much of Hardin. By the time it had receded, billions of dollars of damage had been caused to crops, buildings, and businesses throughout the Midwest. McGowen lost her home and her mother.

Heavy rains in the upper Mississippi and Illinois river valleys started early that year. This portion of the Illinois River reached flood stage in the spring. Because the Illinois from La Grange to Grafton is part of the Alton pool, regulated by the dam on the Mississippi River, it is directly affected by conditions on the bigger river.

After the spring flood receded, the rains came again. Tributaries of both rivers swelled. During one storm in

July, the Spoon River was rising two feet an hour. The water had nowhere to go but up. By early July, Route 100 was closed from Hardin to Pearl, and the ferries were shut down. By late July, levees started to fail, and it was apparent this would be a disaster of epic proportions.

"My mother had been in the hospital, and they moved her to the nursing home in Hardin to recuperate," McGowen is saying. "I was visiting her when they came in and told us East Hardin was being evacuated. They were afraid the levee wouldn't hold."

Her mother had lived in a trailer in East Hardin for thirty years, and McGowen had recently moved in with her. While McGowen and her sisters and brother-in-law took belongings out of the trailer, another crisis arose across the river in Hardin: The nursing home was being evacuated, too.

Her mother was moved to a facility in Hillsboro. Amid the confusion of new patients, no one noticed that her blood sugar was out of control. She died of complications from diabetes.

Six days later the levee broke, covering East Hardin with twenty feet of water. "We walked across the bridge, and you couldn't see the house," she says. "I just couldn't believe it."

In Hardin, the flood inundated the town hall, the bank, businesses, homes, roads, and the nursing home. The cemetery, though, is on high ground. They were able to lay her mother to rest.

"That's one thing we can do here in a flood," *McGowen says. "We can have a funeral."*

THE ROAD HOME

Hardin is the county seat, the largest town in Calhoun County. It is a full-service town, with restaurants, a car dealership, a nursing home, banks, lawyers, a school, a hotel, and taverns.

It grew as a riverboat port and was the primary shipping point for produce grown in the county, apples mostly. Today, most of the orchard goods are shipped out on truck, but a grain elevator keeps the barges coming in. Pleasure boaters and sportsmen put in at the concrete ramp.

Illinois Route 100 crosses to the east at Hardin over the historic Joe Page Bridge. It and the one in Florence are the only vehicular lift bridges on the waterway. Built in 1931, the Joe Page has the lift span of more than 300 feet—almost 100 feet longer than the one in Florence—making it the largest span for a bridge of this type in the world.

The state highway takes travelers through what used to be East Hardin to the bluff line, across Macoupin Creek, and south through the village of Nutwood, past Pere Marquette State Park and into Grafton. Along the way, the rich bottomland—which defines Greene and Scott counties' agricultural riverfront—gets wilder as it narrows.

The morning is fresh and alive with the smell of decay and manure in the air. Queen Anne lace and

Looking upstream toward the Joe Page Bridge from the Jersey County Grain Company's dock area in Hardin. River Mile Marker 21, March 2003.

Before joining with the Mississippi River, the Illinois River's directional path echoes the odd directions of the larger waterway. After curving from a southeasterly direction, the Illinois swings northeast and travels around Gilbert Lake. This small body of water is separated from the river by a narrow strip of land. After this curve, the river runs almost directly east. This change in direction places Grafton, the last downstream community on the Illinois River, on the north side of the river. The photograph was made at sunset looking west up the Illinois River on the right. The Mississippi River is seen beyond the end of the driftwood on the near left side. The island at the extreme left is an unnamed Missouri island that marks the absolute confluence of the two rivers. The island in the center is Island No. 526, also in Missouri. Barely visible in the twilight on the Illinois side is a ferry that runs between Grafton and St. Charles County in Missouri. Illinois River Mile Marker 1 and Mississippi River Mile Marker 219.5, March 2005.

little blue chicory line the roadside, spared, thankfully, from the mower's blades. At the turnoff to the Glades wildlife area, a great blue heron keeps pace with the car as it flies parallel to the roadway.

If at Hardin, though, you stay on the west side of the river, the road leads down the length of Calhoun County, with the sylvan bluffs rising sharply to the right. The town of Brussels rests within those bluffs.

Brussels is a town out of time. It was originally settled in the 1820s, but it wasn't incorporated until 1876. Dubbing itself "a village nestled between rivers," it is about three miles away from the Mississippi River and half a mile off the Illinois.

There's an old hotel and restaurant in town, the Wittmond, which has been in operation continuously since it was established as a trading post in 1847. There are a couple of taverns, a school, a Catholic church, and a cemetery. Across from St. Mary's, Father Tom sits in his van outside the Red and White grocery store and scratches his lottery tickets, placing his hope in one place, his faith in another.

The ferry is about three miles farther away, through the Mark Twain National Wildlife Refuge. It is the final river crossing on the Illinois. Barn swallows swarm about a ferryboat's pilothouse, where mud nests are packed up under the roof. They swoop and follow the ferry back and forth across the river, their homes in constant motion.

The ferry drops you off in Jersey County, almost at the entrance to Pere Marquette State Park, an 8,000-acre preserve with hiking and riding rails, a lodge and cabins, and mounds from the Woodland Indian era.

It's Illinois Route 100 again over here, and it leads south to Grafton.

This is a working river town, belonging as much to the Mississippi River as it does to the Illinois. Fish markets still dot the shoreline, as Grafton stretches out for two miles along the bank. Prone to flooding, it always seems to be in some stage of rebound or decline.

These days the town, founded in 1836, is banking on its historic character. Antique shops and gift shops are thriving and buildings are being restored in town while new, expansive homes are being built atop the bluff. A riverboat festival is held every year. The Old Boatworks—which once churned out paddlewheels and later PT boats during the Korean and Vietnam wars—is a flea market today down at the riverfront, and bed and breakfasts are popping up, too.

One of the gems of Main Street is the Ruebel Hotel and Saloon. After falling into disrepair, the hotel was purchased by Jeff Lorton and his family, who reopened it in 1997. The saloon is more Wild West than river in motif, a testament to Lorton's love of horses and the American West, but despite that incongruity, its windows offer a wide view of the river—or, more precisely, rivers.

"That's the Mississippi River out that window," Lorton says. Then, pointing to the next pane to the right, "And that's the Illinois River over there."

Right here, just outside the Ruebel Hotel in Grafton, is the precise spot where more than 300 years ago, a French cartographer and a priest decided to turn right and head up the Illinois River.

The Interstate 80
Bridge across the Des
Plaines River in Joliet,
photographed from the
east side of the river where
it widens upstream of the
Brandon Road Lock and
Dam. River Mile Marker
286.7, June 2003.

Epilogue

Just north of Grafton, south of the Brussels Ferry, a cross-shaped dolomite monument erected in 1929 commemorates the arrival of Louis Jolliet and Jacques Marquette. It is a bookend with the plaque at the other end, at the Michigan Avenue Bridge in Chicago, the plaque that honors the memory of those explorers who first traveled this waterway.

A trip down the Illinois Waterway reminds us that before those French explorers, this land was home to hundreds of generations of Native Americans. And it reminds us of the glaciers and powerful forces that sculpted this earth before them.

As we look back up that road, we can see all the distinct regions of the Illinois River system: Chicago, Joliet, and the industrial highlands; canal towns, farm towns, and river towns. We see towns whose futures lie in the past. There are richly varied landscapes: hunting grounds and refuges, the straight and narrows, the Pekin Wiggles and the wide Peoria lakes. There are wild areas and levees, locks, and dams.

A look back up that road reminds us of contrasts, too. We see how the river gives life to some towns and wipes others away. We see a way of life fading away and a new generation of opportunities arising.

We see the unique faces of those who live on the river. Each has a voice in the chorus that tells the larger story of the Illinois Waterway. The river can be home, workplace, or playground. It can also be a threat and a scourge. It can be a metaphor. And it can be all too real.

Edgar Lee Masters wrote in *The Sangamon* that "memory is a kind of reading glass under which spots of earth long beloved take on the aspect of something magical" (23). This river is one of those "spots of earth" to a lot of people whose memories and stories are pieces of a whole.

Each of those memories is different. Some people remember the days of deep backwater lakes and skies full of ducks, and some remember the homes they owned before the flood. Some recall bridges or barges or backwater bars. People look at the same thing in different ways, too. A river turtle is a meal to one and a church sermon to another, a symbol of redemption and hope. And depending on who catches it, a carp can be garbage, food, or income. It could be a trophy.

There is a world of difference between the man who fishes from the concrete bank of the Chicago River and the guy running trotlines just north of Browning. They don't look, dress, or talk anything alike. They probably vote differently. Yet they are alike in many ways. They fish the same water. They share a respect for this river. And they both place their faith in an unseen promise that lies just below the surface.

A look back up that road also yields the bigger picture, where the pieces fit together and we recognize the common denominator. We see that it doesn't matter what the fishermen catch—carp or buffalo, catfish, bass, or sauger—it is just a different manifestation of the same dream.

Among those who enjoy and use the river, there is a sense of shared responsibility. The Illinois River is one of the state's greatest natural resources, but it is no longer natural; it's part of an engineered waterway. And because we made it, we cannot escape the responsibility for it. And we have to share.

There is a spot up the river where I pulled off the back road and listened to the insects and the birds. I was up against the bluffs and overlooking the bottomland near Buckhorn Creek. I remember this . . .

The air is fresh here, even on a late summer afternoon. The river is not far. I cannot see the water, but I know it is over there, across the field, beyond the trees.

Silent swallows skim tassels of corn, working the fields of man in search of a meal, and the sparrows call an end of day. The buzz of insects is rhythmic and layered, rising and falling as if it were the breath of a single organism.

At that moment, I hear the drone of diesels, a towboat driving a pack of barges. It comes from downstream, slow and relentless, growing closer and louder until I can see it—the top of the towboat itself—moving beyond the trees. As it passes, the hum of engines changes pitch and slowly fades to the north, gradually giving way to the buzzing of insects and a return to nature, a rhythm of its own.

The sound of that towboat seemed not incongruous with the landscape. The river has a way of assimilating foreigners, of repairing itself. We must give it time. We must be gentle. No matter what we do, though, it will adapt and evolve. And it will make all of us a part of its story.

MAPS   PHOTOGRAPHER'S NOTES   BIBLIOGRAPHY

Photographed on the north side of the Calumet River, between two lift span bridges, on the south side of Chicago. The Indiana Harbor Belt Railroad Bridge and Henry Ford II Memorial Bridge (South Torrence Avenue) are located just east of the Ford Motor Company Chicago Assembly Plant in South Deering. Ford's oldest continually operated facility, production dates back to 1924. Metropolitan Water Reclamation District SEPA Plant Number 1 is located just downstream of the roadway bridge, on the north side of the river, near a six-acre heron rookery. River Mile Marker 328, April, 2007.

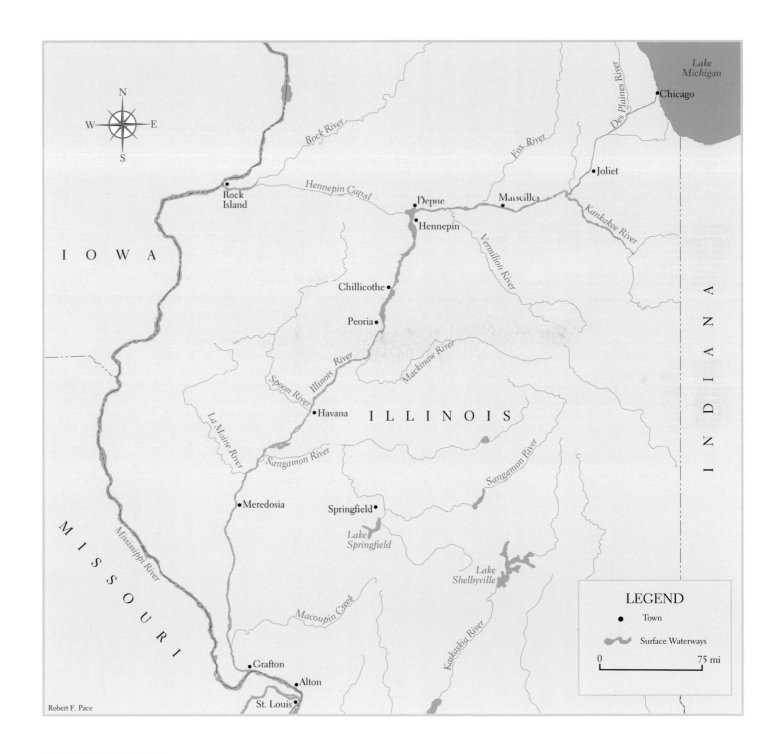

THE ILLINOIS RIVER

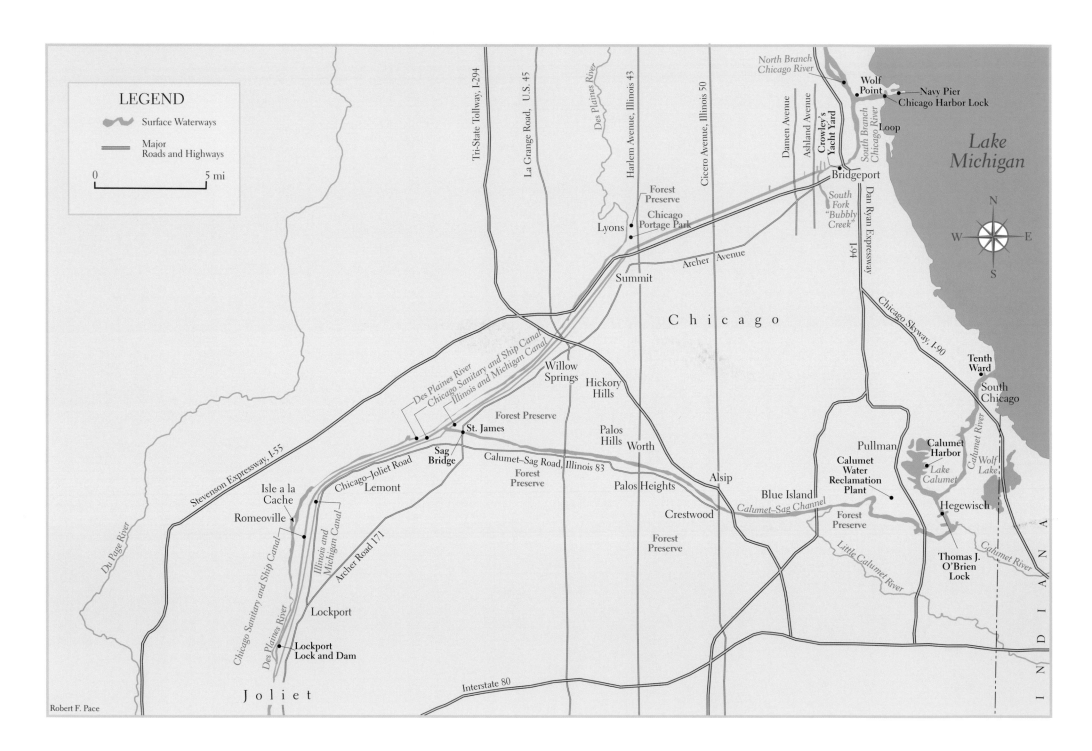

LEGEND

~~~~ Surface Waterways

—— Major Roads and Highways

0 ———— 5 mi

North Branch Chicago River

Wolf Point

Navy Pier
Chicago Harbor Lock

Loop

Lake Michigan

Damen Avenue

Ashland Avenue

Crowley's Yacht Yard

South Branch Chicago River

Bridgeport

South Fork "Bubbly Creek"

Dan Ryan Expressway

I-94

Tri-State Tollway, I-294

La Grange Road, U.S. 45

Des Plaines River

Harlem Avenue, Illinois 43

Cicero Avenue, Illinois 50

Forest Preserve

Chicago Portage Park

Lyons

Summit

Archer Avenue

C h i c a g o

Chicago Skyway, I-90

Tenth Ward

South Chicago

Des Plaines River
Chicago Sanitary and Ship Canal
Illinois and Michigan Canal

Willow Springs

Hickory Hills

Forest Preserve

Palos Hills

Worth

Calumet Harbor

Pullman

Lake Calumet

Wolf Lake

St. James

Sag Bridge

Calumet–Sag Road, Illinois 83

Forest Preserve

Palos Heights

Alsip

Blue Island

Calumet Water Reclamation Plant

Hegewisch

Calumet River

Stevenson Expressway, I-55

Chicago–Joliet Road

Lemont

Isle a la Cache

Romeoville

Archer Road 171

Illinois and Michigan Canal

Crestwood

Calumet–Sag Channel

Forest Preserve

Forest Preserve

Thomas J. O'Brien Lock

Little Calumet River

Calumet River

Du Page River

Chicago Sanitary and Ship Canal

Des Plaines River

Lockport

Lockport Lock and Dam

J o l i e t

Interstate 80

I N D I A N A

Robert F. Pace

THE ILLINOIS WATERWAY, CHICAGO TO JOLIET

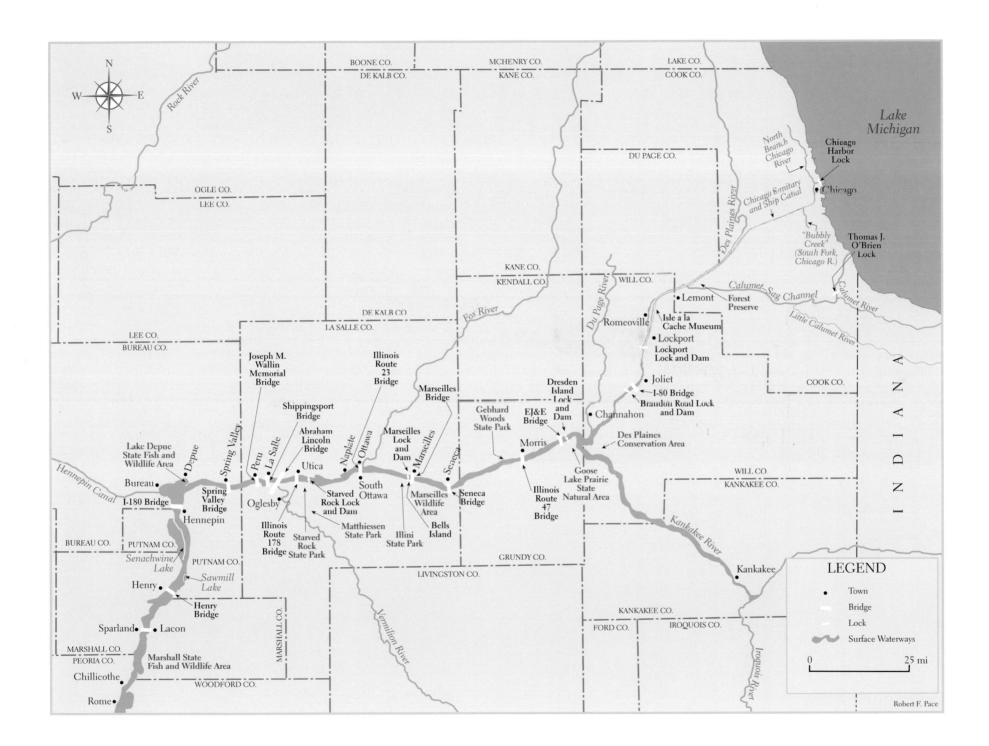

THE UPPER ILLINOIS WATERWAY

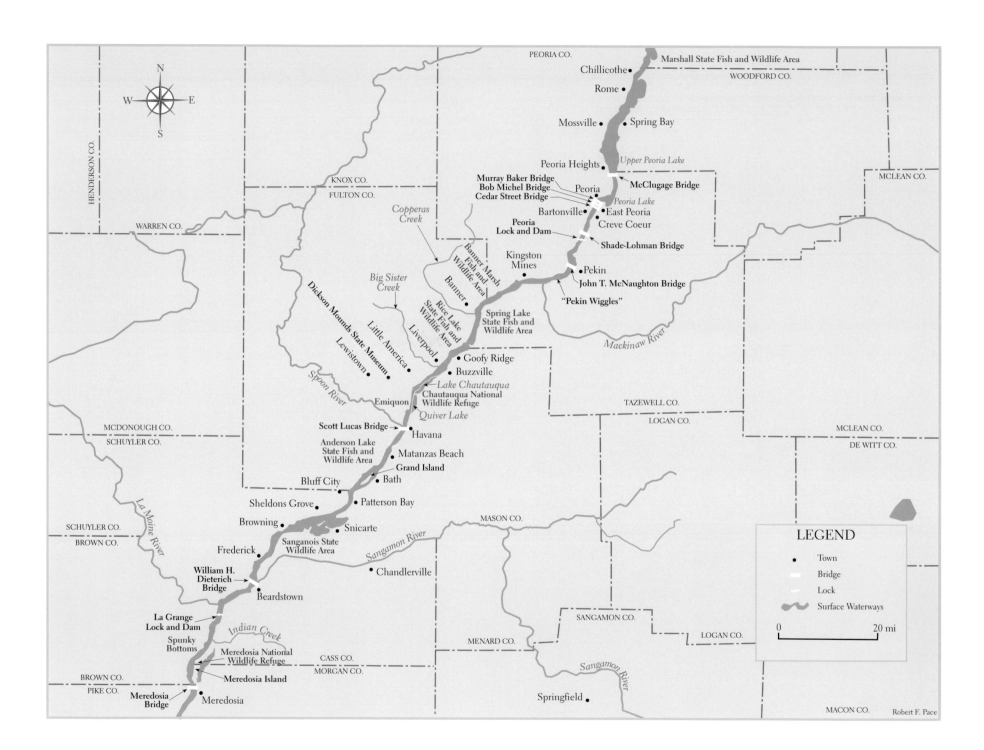

PEORIA CO.

WOODFORD CO.

Marshall State Fish and Wildlife Area

Chillicothe

Rome

Mossville • • Spring Bay

MCLEAN CO.

HENDERSON CO.

KNOX CO.

FULTON CO.

Peoria Heights

Upper Peoria Lake

Murray Baker Bridge
Bob Michel Bridge • Peoria **McClugage Bridge**
Cedar Street Bridge
Peoria Lake
Bartonville • East Peoria
Peoria • Creve Coeur
Lock and Dam
Shade-Lohman Bridge

Copperas Creek

Banner Marsh Fish and Wildlife Area

Kingston Mines

Big Sister Creek

Banner

Pekin
John T. McNaughton Bridge

"Pekin Wiggles"

Rice Lake State Fish and Wildlife Area

Spring Lake State Fish and Wildlife Area

Mackinaw River

Dickson Mounds State Museum

Little America

Liverpool

Lewistown

Goofy Ridge
Buzzville

Spoon River

Lake Chautauqua
Chautauqua National Wildlife Refuge

TAZEWELL CO.

LOGAN CO.

MCLEAN CO.

Emiquon

Quiver Lake

Scott Lucas Bridge → Havana

DE WITT CO.

MCDONOUGH CO.

SCHUYLER CO.

Anderson Lake State Fish and Wildlife Area

Matanzas Beach

Grand Island

Bluff City • Bath

Sheldons Grove • Patterson Bay

MASON CO.

SCHUYLER CO.

BROWN CO.

La Moine River

Browning

Snicarte

Sangamon River

Frederick

Sanganois State Wildlife Area

Chandlerville

LEGEND

• Town

**William H.
Dieterich
Bridge**

Beardstown

SANGAMON CO.

LOGAN CO.

▭ Bridge

┄ Lock

〰 Surface Waterways

0   20 mi

**La Grange
Lock and Dam**

Indian Creek

Spunky
Bottoms

Meredosia National
Wildlife Refuge

CASS CO.

MORGAN CO.

MENARD CO.

Sangamon River

Meredosia Island

BROWN CO.

**Meredosia
Bridge** → Meredosia

Springfield •

PIKE CO.

MACON CO.

Robert F. Pace

THE MIDDLE ILLINOIS WATERWAY

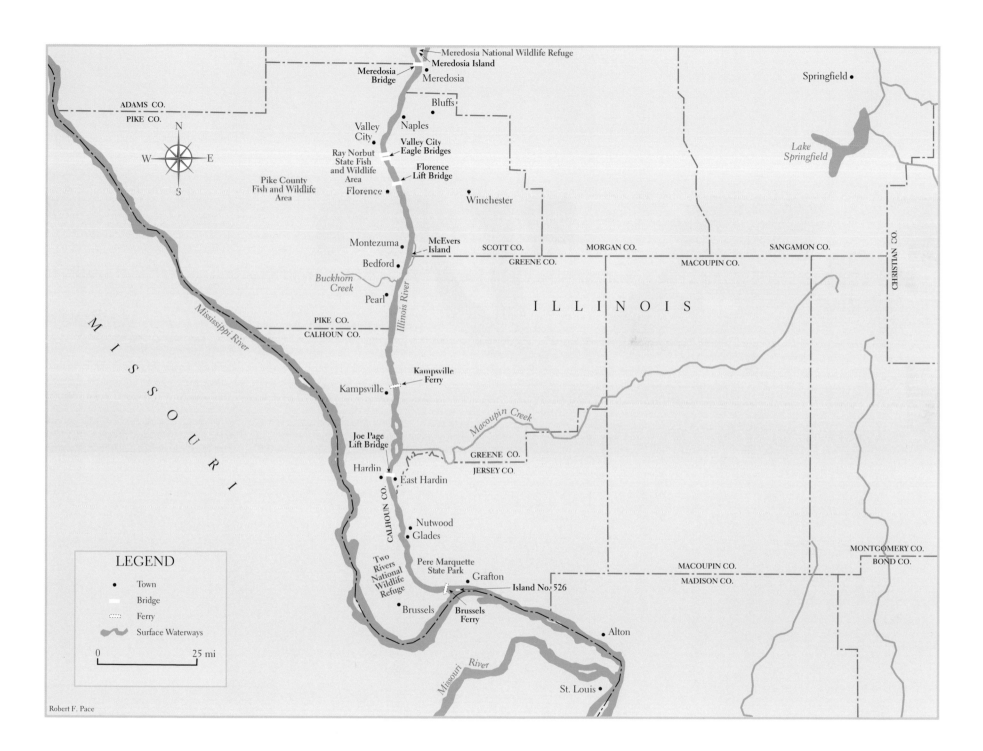

Meredosia National Wildlife Refuge
Meredosia Island
Meredosia Bridge
Meredosia

Bluffs

Naples

Valley City
Valley City Eagle Bridges
Ray Norbut State Fish and Wildlife Area
Florence Lift Bridge
Florence

Pike County Fish and Wildlife Area

Winchester

McEvers Island

Montezuma

SCOTT CO.
MORGAN CO.
SANGAMON CO.
GREENE CO.
MACOUPIN CO.

Bedford

Buckhorn Creek

Pearl

PIKE CO.
CALHOUN CO.

Illinois River

I L L I N O I S

Mississippi River

M I S S O U R I

Springfield

Lake Springfield

CHRISTIAN CO.

Macoupin Creek

GREENE CO.
JERSEY CO.

Kampsville Ferry
Kampsville

Joe Page Lift Bridge
Hardin
East Hardin

CALHOUN CO.

Nutwood
Glades

MONTGOMERY CO.
BOND CO.

Two Rivers National Wildlife Refuge

Pere Marquette State Park
Grafton
Island No. 526

MACOUPIN CO.
MADISON CO.

Brussels
Brussels Ferry

Alton

Missouri River

St. Louis

ADAMS CO.
PIKE CO.

N
W E
S

LEGEND

• Town
▬ Bridge
┈ Ferry
〰 Surface Waterways

0 _____ 25 mi

Robert F. Pace

*The 2004 PRO Nationals boat races on
Depue Lake in Depue, located just north of
the Illinois River, in a 360-degree view, center
facing southeast. Near River Mile Marker
211, July 2004.*

PHOTOGRAPHER'S NOTES

The photographs in this book were taken over an eight-year period, dating back to 1999. Because there are different formats and methods utilized in various images, I thought a brief note about the making of the work might be of interest.

All of the images were made on Kodak (120) and Fuji (35mm) color negative film. The film was scanned using a variety of film scanners to produce the files for the book. Only minor corrections, mainly for dust spotting, were made to the original images in Photoshop for prepress preparation. (The Morris Bridge destruction sequence is the exception. Original images were 35mm cropped to pan.) The formats, both square and panoramic, are the products of the original cameras used in making the pictures.

There were a number of cameras employed:

Linhof 617 IIIs, 72mm and 180mm lenses
Hasselblad SWC/M
Hasselblad ArcBody, 35mm lens
Hasselblad Xpan, 45mm and 90mm lenses
Hasselblad EL/M, 60mm lens (mounted on a Kenyon
     KS-4 GyroStabilizer for aerial images)
Roundshot Super VR 220 with a PC-Nikkon 35mm lens
Nikon F4, 105mm lens (Morris Bridge destruction
     sequence)

A Balcar Concept B3 portable flash was used in many of the portraits. A Quantum 400-watt Q-Flash unit was used as well.

The Sand Point Light and Daymark at sunrise,
looking east from Rome. River Mile Marker
176.6, April 2005.

BIBLIOGRAPHY

Easley, Courtney A. "The Way I Saw It." Meredosia, Ill.: self-published, 2002.

Esarey, Duane. "Emiquon, a Place in Nature, a Place in Time." *Living Museum* 60, nos. 1 and 2 (Winter 1997–98; Spring/Summer 1998).

Gray, James. *The Illinois*. 1940. Reprint, Urbana: University of Illinois Press, 1989.

Hamm, Dale, with David Bakke. *The Last of the Market Hunters*. Carbondale: Southern Illinois University Press, 1996.

Havera, Stephen P. *Waterfowl of Illinois. Status and Management*. Urbana: Illinois Natural History Survey, 1999.

Higgins, Kenneth F. *Childhood Memories from the West Bank of the Illinois River*. Sioux Falls, S.D.: Pine Hill Press, 2004.

Hill, Libby. *The Chicago River: A Natural and Unnatural History*. Chicago: Lake Claremont Press, 2000.

Lamb, John. *A Corridor in Time*. Romeoville, Ill.: Lewis University in cooperation with the Illinois Department of Commerce and Community Affairs Office of Tourism, 1987.

Marlin, John C. "Long Distance Transport of Illinois River Dredged Material for Beneficial Use in Chicago." Proceedings of the Western Dredging Association Twenty-fourth Technical Conference and Thirty-sixth Texas A&M Dredging Seminar, Orlando, Fla., July 6–9, 2004. <http://www.wmrc.uiuc.edu/special_projects/il_river/long-distance-il-river-sediment-transport.pdf>.

Masters, Edgar Lee. *The Sangamon*. 1942. Reprint, Prairie State Books ed. Urbana: University of Illinois Press, 1988.

——. *Spoon River Anthology*. 1915. Reprint, New York: Signet Classic, New American Library, Penguin Putnam, 1992.

Matson, Nehemiah. *French and Indians of Illinois River*. 1874. Reprint, Shawnee Classics ed. Carbondale: Southern Illinois University Press, 2001.

Meredosia Bicentennial Book 1776–1976. Bluffs, Ill.: Jones Publishing Company, 1976.

Morgan, Lael. *Good Time Girls of the Alaska-Yukon Gold Rush*. Fairbanks, Alaska: Epicenter Press, 1998.

Ranney, Edward, and Emily J. Harris. *Prairie Passage: The Illinois and Michigan Canal Corridor*. Urbana: University of Illinois Press, 1998.

Slosowski, Donna. *St. James at Sag Bridge Church, History and Cookbook*. Lemont, Ill.: St. James Ladies Guild (self-published), n.d.

Solzman, David M. *The Chicago River: An Illustrated History and Guide to the River and Its Waterways*. Chicago: Wild Onion Books, Loyola Press, 1998.

Soong, David Ta Wei. Post-workshop summary. The Sino–U.S. Joint Workshop on Sediment Transport and Sediment Induced Disasters, Beijing, China,

March 15–17, 1999. Prepared for the National Science Foundation. <http://www.sws.uiuc.edu/pubdoc/IEM/ISWSIEM2000-01.pdf>.

Thompson, John. *Wetlands Drainage, River Modifications, and Sectoral Conflict in the Lower Illinois Valley, 1890–1930.* 1924. Reprint, Carbondale: Southern Illinois University Press, 2002.

Twain, Mark. *Life on the Mississippi.* In *Mississippi Writings.* New York: Literary Classics of the United States, 1982.

Young, David M. *Chicago Maritime.* De Kalb: Northern Illinois University Press, 2001.

DANIEL OVERTURF is an associate professor in the Department of Cinema and Photography at Southern Illinois University Carbondale. Born and raised in Peoria, he received degrees from SIUC in photography and printmaking. Prior to returning to SIUC to teach in 1990, he worked as a photographer and teacher in New Mexico, Kansas, Nevada, and Alberta.

GARY MARX is a copy editor for the *Kansas City Star* and a freelance writer. He grew up near the Des Plaines River in Schiller Park. He holds a journalism degree from Southern Illinois University Carbondale and has worked as a reporter, editor, and columnist at newspapers in Indiana, Illinois, and Missouri.

A River Through Illinois

Designed and typeset by Erin New

Printed and bound by Four Colour Imports, Ltd.

Composed in Electra

Printed on 130 gsm Lumi Silk

Bound in Saifu cloth